I0212767

Other works illustrated, annotated and formatted by John W. Cirignani include:

My Life on the Plains

by George A. Custer

and

A COMPLETE LIFE OF GENERAL GEORGE A. CUSTER

MAJOR-GENERAL OF VOLUNTEERS; BREVET MAJOR-GENERAL, U.S. ARMY; AND LIEUTENANT-COLONEL, SEVENTH U.S. CAVALRY

By
Frederick Whittaker

www.HistoryInWordsAndPictures.com

HISTORY IN WORDS AND PICTURES SERIES

BOOK FIVE

PRESENTED BY JOHN W. CIRIGNANI

PERSONAL RECOLLECTIONS OF A CAVALRYMAN WITH CUSTER'S MICHIGAN CAVALRY BRIGADE
IN THE CIVIL WAR

By

J.H. Kidd

FORMERLY COLONEL SIXTH MICHIGAN CAVALRY
AND
BREVET BRIGADIER GENERAL OF VOLUNTEERS

Copyright © Entered according to Act of Congress, in the year
1908, by
SENTINEL PRINTING COMPANY, Ionia, Michigan
In the Office of the Librarian of Congress at Washington

Formatted Version, Illustration Order and Author Illustrations
by John W. Cirignani.
Copyright © John Cirignani, 2015
All Rights Reserved

ISBN: 978-0-9966994-2-6

Illustration Credits located at book end.

TO MY WIFE AND SON
AND
TO MY COMRADES OF THE MICHIGAN CAVALRY
BRIGADE
This volume is affectionately dedicated – J.H. Kidd

FOREWORD

The *History in Words and Pictures Series* is a collection of books written at the time the history occurred, by the people who made it. We often "learn" our history from one author's interpretation, which then becomes an interpretation of an interpretation, then morphs several more times over generations until the interpretations are accepted as fact.

I found this especially true when researching General George Armstrong Custer. I had been developing a screenplay for a major Hollywood movie producer, the topic which included Custer. After reading his own words in *My Life on the Plains*, I started to come to the conclusion Custer was not the man typically depicted in various books and movies. Further research confirmed my suspicions. The Custer stereotype is so different from the actual man that I began to wonder how this phenomenon began.

The answer is that we tend to rely on interpretations of authors who, whatever their intentions, see history through their own contemporary biased lens. What makes us rely on this material? The answer is simple; the interpretation is packaged in a format that we are used to.

The primary problem in reading historical works is the antiquated reading experience. Single paragraphs often stretch over several pages. Punctuation is different. Photographs and illustrations are rare.

Therefore, rather than rewrite history, I've reformatted the text to modern standards. I added annotated illustrations and photography, contemporary when possible, of the people and places included in the story. The result is a visual characteristic absent from most works published in the 19th century and prior. The book is then packaged into a series that will continue so long as I find interesting material.

PERSONAL RECOLLECTIONS OF A CAVALRYMAN WITH CUSTER'S MICHIGAN CAVALRY BRIGADE is the fifth book of the series. The sixth book in the series book was written by Custer's official biographer, Frederick Whittaker. I'll add new works to the collection on a regular basis.

My wish is that the *History in Words and Pictures Series* will help ignite a new interest in the past that doesn't exist among our younger generations. There is so much fascinating material, and so much to be learned from it. How it is interpreted is still up to the individual, but at least they're getting it from "the horse's mouth."

Finally, as you read *Boots and Saddles,* imagine James Kidd himself sitting across the room, legs crossed, reading to you. After all, it is he who is speaking!

John W. Cirignani

CONTENTS

PREFACE

IN preparing this book, it has not been the purpose of the author to write a complete historical sketch of the Michigan Cavalry Brigade. Such a history would require a volume as large for the record of each regiment; and even then, it would fall short of doing justice to the patriotic services of that superb organization. The narrative contained in the following pages is a story of the personal recollections of one of the troopers who rode with Custer, and played a part - small it is true, but still a part - in the tragedy of the Civil War. As such it is modestly put forth, with the hope that it may prove to be an interesting story to those who read it. The author also trusts that it may contribute something, albeit but a little, toward giving Custer's Michigan cavalrymen the place in the history of their country which they so richly earned on many fields.

Doubtless, many things have been omitted that ought to have been included, and some things written in that it might have been better to leave out. These are matters of personal judgment and taste, and no man's judgment is infallible. The chapters have been written in intervals of leisure during a period of more than twenty years. The one on Cedar Creek appeared first in 1886; the Gettysburg campaign in 1889; Brandy Station, Kilpatrick's Richmond expedition, the Yellow Tavern campaign, Buckland Mills, Hanovertown and Haw's Shop, The Trevilian Raid and some other portions have been prepared during the current year - 1908.

While memory has been the principal guide, the strict historical truth has been sought, and when there appeared to be a reasonable doubt, the official records have been consulted and the writings of others freely drawn upon to verify these recollections.

The Memoirs of P.H. Sheridan and H.B. McClellan's Campaigns of Stuart's Cavalry have been of especial value in this respect; the latter helping to give both sides of the picture, particularly in the accounts of the battles of Buckland Mills and Yellow Tavern. Wade Hampton's official reports were put to similar use in describing the Battle of Trevilian Station.

So far as mention is made of individual officers and men, there is no pretense that the list is complete. Those whose names appear in the text were selected as types. Hundreds of others were equally deserving. The same remark applies to the portraits. These are representative faces. The list could be extended indefinitely.

It was intended to include in an appendix a full roster of all the men who served in the Sixth Michigan Cavalry and in the other regiments as well; but this would have made the book too bulky. By applying to the adjutant general of Michigan, the books published by the state giving the record of every man who served in either of the regiments in the brigade can be obtained.

The Roll of Honor - a list of all those who were killed in action, or who died of wounds received in action - is as complete as it was possible to make it from the official records. In a very few cases, men who were reported missing in action, and of whom no further record could be found, were assumed to have belonged in the list, but these are not numerous enough to materially affect the totals.

For the rest, the author cannot claim that he has done justice to either of these organizations, but he has made an honest effort to be fair and impartial, to tell the truth as he saw it, without prejudice. How well he has succeeded is not for him to say. "It is an interesting story," said an officer who served with distinction in the Fifth Michigan Cavalry. If that shall be the verdict of all the comrades who read it, the writer will be satisfied.

A NATIONAL AWAKENING

THE war cloud that burst upon the country in 1861 was no surprise to sagacious observers. For many years it had been visible, at times a mere speck in the sky, again growing larger and more angry in appearance. It would disappear, sanguine patriots hoped forever, only to come again, full of dire portent and evil menacings. All men who were not blind saw it, but most of them trusted, many believed, that it would pass over and do no harm. Some of those high in authority blindly pinned their faith to luck and shut their eyes to the peril. Danger signals were set, but the mariners who were trying to steer the Ship of State let her drift, making slight, if any, efforts to put her up against the wind and keep her off the rocks.

It is likely, however, that the Civil War was one of those things that had to be; that it was a means used by destiny to shape our ends; that it was needed to bring out those fine traits of national character which, up to that time, were not known to exist. Southern blood was hot and Northern blood was cold. Though citizens of one country, the people of the North and the people of the South were separated by a wide gulf in their interests and in their feelings. Doubt had been freely thrown upon the courage of the men who lived north of Mason and Dixon's line. The haughty slave owners and slave dealers affected to believe, many of them did believe, that one southern man could whip five "Yankees." It took four years of war to teach them a different lesson.

It was the old story of highland and lowland feud, of the white rose and the red rose, of roundhead and cavalier, of foemen worthy of each other's steel fighting to weld discordant and belligerent elements into a homogeneous whole.

But war is not always an unmixed evil. Sometimes it is a positive good, and the nation emerged from its great struggle more united than ever. The sections had learned to respect each other's prowess and to know each other's virtues. The cement that bound the union of states was no longer like wax to be melted by the fervent heat of political strife. It had been tested and tempered in the fiery furnace of Civil War. The history of that war often has been written. Much has been written that is not history. But whether fact or fiction, the story is read with undiminished interest as the years rush by.

One story there is that has not been told, at least not all of it; nor will it be until the last of those who took part in that great drama shall have gone over to the silent majority. It is the story of the individual experiences of the men who stood in the ranks, or of the officers who held no high rank; who knew little of plans and strategy, but bore their part of the burden and obeyed orders. There was no army, no corps, no division, brigade or regiment, scarcely a battery, troop or company, which went through that struggle, or a soldier who served in the field for three years or during the war, whose experiences did not differ from any other, whose history would not contain many features peculiar to itself or himself.

Two regiments in the same command, two soldiers in the same regiment, might get entirely different impressions of the battle in which both participated. Two equally truthful accounts might vary greatly in their details. What one saw, another might not see, and each could judge correctly only of what he himself witnessed. This fact accounts, in part, for the many contradictions, which are not contradictions, in the annals of the war. The witnesses did not occupy the same standpoint. They were looking at different parts of the same panorama. Oftentimes they are like the two knights who slew each other in a quarrel about the color of a shield. One said it was red; the other declared it was green. Both were right, for it was red on one side and green on the other.

On such flimsy pretexts do men and nations wage war. Why then wonder if historians differ also? In the "Wilderness," each man's view was bounded by a very narrow horizon, and few knew what was going on outside their range of vision. What was true of the Wilderness was true of nearly every battle fought between the Union and Confederate forces. No picture of a battle, whether it be painted in words or in colors, can bring into the perspective more than a glimpse of the actual field. No man could possibly have been stationed where he could see it all.

Hence it came to pass that many a private soldier knew things which the corps commander did not know; and saw things which others did not see. The official reports, for the most part, furnish but a bare outline, and are often misleading. The details may be put in by an infinite number of hands, and those features that seen separately appear incongruous, when blended will form a perfect picture. But it must be seen, like a panorama, in parts, for no single eye could take in at once all the details in a picture of a battle.

In the winter of 1855-56, while engaged as assistant factotum in a general lumbering and mercantile business in the pine woods of Northern Michigan, one of my functions was that of assistant postmaster, which led to getting up a "club" for the New York Weekly Tribune, the premium for which was an extra copy for myself. The result was that in due time, my mind was imbued with the principles of Horace Greeley. The boys who read the Tribune in the fifties were being unconsciously molded into the men who a few years later rushed to the rescue of their country's flag. The seed sown by Horace Greeley and others like him brought forth a rich crop of loyalty, of devotion and self-sacrifice that was garnered in the war.

In the latter part of the year 1860, the air was full of threatenings. The country was clearly on the verge of civil war, and the feeling almost as intense as it was in the following April, after the flash of Edmund Ruffin's gun had fired the Northern heart.

In October, I came a freshman into the University of Michigan in Ann Arbor. That noble institution was, even then, the pride of the Peninsula state. A superb corps of instructors, headed by Henry P. Tappan, the noblest Roman of them all, smoothed the pathway to learning which a thousand young men were trying to tread. These boys were full of life, vigor, ambition and energy. They were from various parts of the country, though but few were from the Southern States. The atmosphere of the place was wholesome, and calculated to develop a robust, courageous manhood. The students were led to study the best antique models, and to emulate the heroic traits of character in the great men of modern times. It may be said that nowhere in the land did the fires of patriotism burn with more fervent heat during the eventful and exciting period that preceded by a few months the inauguration of Abraham Lincoln.

The University of Michigan was founded in 1817 as the Catholepistemiad, or University of Michigania. The school moved from Detroit to Ann Arbor in 1837. The first classes were held in 1841, and eleven men graduated in the first commencement ceremony of 1845.

The young men took a deep interest in the political campaign of that year, and watched with eager faces for every item of news that pertained to it. The nomination of Abraham Lincoln was a bitter disappointment to the young Republicans of Michigan. Seward was their idol and their ideal, and when the news came of his defeat in the Chicago Convention, many men shed tears who later learned to love the very ground on which the Illinois "Railsplitter" stood; and who today cherish his memory with the same reverential respect which they feel for that of Washington.

During that memorable campaign, Seward spoke in Detroit and scores of students went from Ann Arbor to hear him. He did not impress one as a great orator. He was of slight frame, but of a noble and intellectual cast of countenance. His arguments were convincing, his language well chosen, but he was somewhat lacking in the physical attributes so essential to perfect success as a public speaker. His features were very marked, with a big nose, a firm jaw, a lofty forehead and a skin almost colorless. He had been the choice of Michigan for president, and was received with the warmest demonstrations of respect and enthusiasm. Every word that fell from his lips was eagerly caught up by the great multitude. It was a proud day for him, and his heart must have been touched by the abounding evidences of affection.

Seward was looked upon as the embodiment of sagacious statesmanship and political prescience, but how far he fell short of comprehending the real magnitude of the crisis then impending was shown by his prediction that the war would last but ninety days. His famous dictum about the "irrepressible conflict" did him more credit.

That same year, Salmon P. Chase also spoke in Michigan. There were giants in those days. Chase was not at all like Seward in his appearance. Tall and of commanding figure, he was a man of perfect physique. He had an expressive face and an excellent voice, well adapted to outdoor speaking. In manner, he appeared somewhat pompous, and the impression he left on the mind of the listener was not so agreeable as that retained of the great New Yorker.

A crowd gathered in front of the post office on Griswold Street in Detroit, Michigan, for the administering of the Oath of Allegiance. In the distance is the Baptist Church Tower, now torn down. The date is April 20, 1861.

At some time during the summer of 1860, Stephen A. Douglas passed through Michigan over the Central Railroad. His train stopped at all stations and hundreds of students flocked to see and hear him. He came off the car to a temporary platform, and for twenty minutes, that sea of faces gazing at him with rapt attention, talked with great rapidity, but with such earnestness and force as to enchain the minds of his hearers. His remarks were in part stereotyped, and he made much of his well worn argument about the right of the territories to "regulate their own domestic institutions in their own way, subject only to the constitution."

In manner, he was easy and graceful; in appearance, striking. He spoke with no apparent effort. Of massive frame though short in stature after the manner of General Sheridan, his head was large and set off by a luxuriant growth of hair that served to enhance its apparent size. His face was smooth, full and florid, the hue rather suggestive. His countenance and bearing indicated force, courage and tenacity of purpose. I was not surprised when he announced that he was on the side of the Union, and believe that had he lived, he would have been, like Logan, a great soldier and a loyal supporter of Lincoln. He was a patriot of the purest type, and one of the ablest men of his time.

Stephen Douglas was the Democratic Party nominee for President in the 1860 election, losing to Republican Abraham Lincoln. Douglas had previously defeated Lincoln in a Senate race, noted for the famous Lincoln-Douglas debates of 1858.

A significant incident of the winter of 1860-61 seems worth recalling. That period was one of the most intense excitement. What with the secession of the Southern States, the resignation of Senators and Members of Congress, and the vacillating course of the Buchanan administration, the outlook was gloomy in the extreme. There were in the University a number of students from the South, and they kept their trunks packed, ready to leave at a moment's notice. Party feeling ran high, and the tension was painful.

As the publisher of The Liberator, a fiery anti-slavery newspaper, Garrison was at the forefront of the crusade against slavery from the 1830s until he felt the issue had been settled by the passage of the 13th Amendment following the Civil War.

William Lloyd Garrison came to Ann Arbor to speak and could not get a hall, but finally succeeded in securing a building used for a school house in the lower part of the town. Here he was, set upon by a lot of roughs who interrupted him with cat calls and hisses, and made demonstrations so threatening that to avoid bodily injury, he was compelled to make his exit through a window. The affair was laid to the students, and some of them were engaged in it, to their discredit, be it said. It was not safe for an Abolitionist to free his mind even in the "Athens" of Michigan.

Harper's Weekly published an illustrative cut of the scene, and Ann Arbor achieved an unenviable notoriety.

Henry Tappan's speech at Ann Arbor Court House Square, announcing that Fort Sumter had been fired upon. Henry Philip Tappan was the first president of the University of Michigan. He was instrumental in fashioning the university as a prototype for American research universities. Tappan was convinced of the superiority of the German model of public education, with a system of primary schools, secondary schools, and a university, all administered by the state and supported with tax dollars.

One day, all hands went to the train to see the Prince of Wales, who was to pass through on his way to Chicago. There was much curiosity to see the queen's son. He had been treated with distinguished consideration in the East and was going to take a look at the Western metropolis. There was a big crowd at the station, but his royal highness did not deign to notice us, much less to come out and make a speech as Douglas did, who was a much greater man. But the "Little Giant" was neither a prince nor the son of a prince, though a "sovereign" in his own right, as is every American citizen. Through the open window however, we had a glimpse of the scion of royalty, and saw a rather unpretentious looking young person in the garb of a gentleman. The Duke of Newcastle stood on the platform, where he could be seen, and looked and acted much like an ordinary mortal. The boys agreed that he might make a very fair governor or congressman, if he were to turn Democrat and become a citizen of the land of the free and the home of the brave.

The faculty in the University of Michigan in 1860 was a brilliant one, including the names of many who have had a worldwide reputation as scholars and savants. Andrew D. White, since President of Cornell University and distinguished in the diplomatic service of his country, was professor of history. Henry P. Tappan, President of the University, or Chancellor, as he was fond of being styled after the manner of the Germans, was a magnificent specimen of manhood, intellectually and physically. Tall and majestic in appearance, he had a massive head and noble countenance, an intellect profound and brilliant. No wonder that he was worshiped, for he was god-like in form and in mind. Like many another great man, however, it was his fate to incur the enmity of certain others too narrow and mean to appreciate either his ability or his nobility of character. Being on the Board of Regents they had the power, and used it relentlessly, to drive him out of the seat of learning which he had done more than all others to build up and to honor.

The University was his pride and glory, and when he was thus smitten in the house of his friends, he shook the dust from his feet and went away, never to return. It is a sad story. He died abroad, after having been for many years an exile from his native land. The feeling against these men was bitter in the extreme. The students hung one of them in effigy and marched in a body to the house of the other, and assailed it with stones and missiles, meantime filling the air with execrations on his head. Both long since ceased to be remembered even by name, but the memory of Tappan remains as one of the choicest traditions of the University, and it will be as enduring as the life of the institution itself.

AN EVENTFUL WINTER

IT was an eventful winter that preceded the breaking out of the War Between the States. The salient feature of the time, apart from the excitement, was the uncertainty. War seemed inevitable, yet the temporizing continued. The South went on seizing forts and plundering arsenals, terrorizing Union sentiment and threatening the Federal government. The arming of troops proceeded without check, and hostile cannon were defiantly pointed at Federal forts. Every friend of his country felt his cheek burn with shame, and longed for one day of Andrew Jackson to stifle the conspiracy while it was in its infancy.

One by one the states went out, boldly proclaiming that they owed no allegiance to the government; but the leaders in the North clung to the delusion that the bridges were not all burned and that the erring ones might be coaxed or cajoled into returning. Concessions were offered, point after point was yielded, even to the verge of dishonor, in an idle attempt to patch up a peace that from the nature of the case could have been but temporary, if obtained on such terms. The people of the Northern States had set their faces resolutely against secession, and led by Lincoln had crossed the Rubicon and taken up the gage of battle which had been thrown down by the South.

There was then no alternative but to fight. All other schemes were illusive. The supreme crisis of the nation had come, and there was no other way than for the loyalty of the country to assert itself. The courage of the people had to be put to the proof, to see whether they were worthy of the heritage of freedom that had been earned by the blood of the fathers. For fifty years, there had been no war in this country except the affair with Mexico, so far away that distance lent enchantment to the view. The Northern people had not been bred to arms. The martial spirit was well-nigh extinct. Men knew little of military exercises, except such ideas as had been derived from the old militia system that in many states was treated by the people rather with derision than respect, and in most of them was, in the impending emergency, a rather poor reliance for the national defense. Southerners, trained in the use of firearms and to the duello, did not attempt to conceal their contempt for their Northern brethren, and feigned to believe that north of Mason and Dixon's line lived a race of cowards.

It did not take long to demonstrate that the descendants of the Green Mountain Boys and of the western pioneers were foes worthy of the mettle of the men who came from the states of Sumter and Marion, and "Light Horse Harry Lee." The blood of their heroic ancestry ran in their veins, and they were ready and willing to do or die when once convinced that their country was in deadly peril. The people, indeed, were ready long before their leaders were. Some of the ablest men the North had produced were awed by their fear of the South - not physical fear, for Webster and Douglas and Cass were incapable of such a thing - but fear that the weight of Southern political influence might be thrown against them. Many of the party leaders of the North had come to be known as "dough-faces," a term of reproach, referring to the supposed ease with which they might be kneaded into any form required for Southern use. They might have been styled very appropriately "wax-nosed politicians," after the English custom, from the way they were nosed around by arrogant champions of the cause of slavery.

Conciliation was tried, but every effort in that direction failed. A tempest of discussion arose over the "Crittenden Compromise

Resolutions," the last overture for peace on the part of the North. It was generally conceded that it would be better to have war than to give up all for which the North had been contending for so many years. There was a feeling of profound indignation and disgust at Buchanan's message to Congress, in which he virtually conceded the right of secession and denied the power of the federal government to coerce a state.

Governor Austin Blair was sworn into office just two weeks after South Carolina seceded from the Union. He was a strong opponent of slavery, he championed the effort to ban capital punishment and he supported efforts to give women and black citizens the right to vote.

The course of General Cass in resigning from the Cabinet, rather than be a party to the feeble policy of the President, was applauded by all parties in Michigan, and the venerable statesman resumed his old-time place in the affections of the people of the Peninsula state. Governor Blair voiced the sentiments of Democrats and Republicans alike when he practically tendered the whole power of the state to sustain the federal government in its determination to maintain the Union. All the utterances of the "War Governor" during that trying period breathed a spirit of devoted patriotism and lofty courage. The people were with him, and long before the call to arms was sounded by President Lincoln, the "Wolverines" were ready to do their part in the coming struggle.

In the evening of the day when Fort Sumter was fired upon, the students marched in a body to the house of Chancellor Tappan and called him out. His remarks were an exhortation to duty, an appeal to patriotism. He advised against haste, saying that the chances were that the country would be more in need of men in a year from that time than it was then. The University would put no hindrance in the way of such students as might feel impelled by a sense of duty to respond to the call for troops, but on the contrary, would bid them Godspeed and watch their careers with pride and solicitude. The speech was calm, but filled with the loftiest sentiments.

Professor Andrew D. White was also visited and made a most memorable and significant speech. Standing on the porch of his house in the presence of several hundred young men, he declared his opinion that one of the greatest wars of history was upon us; which he believed would not end in a day, but would be a protracted and bloody struggle. "I shall not be surprised," said he, "if it turns out to be another Thirty Years War, and no prophet can predict what momentous consequences may result from it, before a Gustavus Adolphus shall arise to lead the armies of the Union to victory." He made a rousing Union speech that was loudly cheered by the throng of young men who heard it. Dr. Tappan also addressed an immense mass meeting, and all things worked together to arouse the entire people to a high pitch of enthusiastic ardor for the cause of the Union.

At once, the town took on a military air. The state militia companies made haste to respond to the first call for three months' service, and were assigned to the First Regiment of Michigan Infantry, stationed in Detroit. The ranks were filled to the maximum in an incredibly short space of time. Indeed, there were more men than munitions for the service, and it was more difficult to equip the troops than to enlist them. The position of private in the ranks was much sought.

As an illustration of this: On the afternoon before the First Regiment of Michigan three-months men was to leave Detroit to march to Washington, my roommate, William Channing Moore, a member of the Freshman class, came hurriedly into the room and aglow with excitement, threw down his books, and extending his hand said; "Good-by, old boy; there is a vacant position in the Adrian Company. I have accepted it and am off for the war. I leave on the first train for Detroit and shall join the company tomorrow morning." "What is the position?" I asked. "High private in the rear rank," he laughingly replied.

Moore was in the Bull Run battle, where he was shot through the arm and taken prisoner. He was exchanged and discharged and came back to his class in 1862. His sense of duty was not satisfied however, for he enlisted again in the Eighteenth Michigan Infantry, in which regiment he rose to be a captain. He survived the war and returned to civil life, only to be drowned several years later while fording a river in the South. "Billy" Moore, as he was affectionately called, was a young man of superb physique, an athlete, a fine student and as innocent of guile as a child. He is mentioned here as a typical student volunteer, one of many, as the record of the Michigan University in the war amply proves.

Two other University men worthy to be named in the list with Moore were Henry B. Landon and Allen A. Zacharias. Landon was graduated from the literary department in 1861. He immediately entered service as adjutant of the Seventh Michigan infantry - the regiment which led the advance of Burnside's army across the river in the Battle of Fredericksburg. He was shot through the body in the Battle of Fair Oaks, the bullet it was said, passing through both lungs. This wound led to his discharge for disability. Landon returned to Ann Arbor and took a course in the medical department of the University, after which he reentered service as assistant surgeon of his old regiment. He survived the war, and became a physician and surgeon of repute, a pillar in the Episcopal Church and an excellent citizen. Landon was a prince of good fellows, always bubbling over with fun, drollery and wit; and withal, a fine vocalist with a rich bass voice. In the winter of 1863-64, he often came to see me in my camp on the Rapidan, near Stevensburg, Virginia, and there was no man in the army whose visits were more welcome.

Andrew White White was a Professor of History and English literature at the University of Michigan from 1858 until 1863. White made his lasting mark on the grounds of the university by enrolling students to plant elms along the walkways on The Diag, a large open space in the middle of the university's Central Campus.

Zacharias was graduated in 1860. He went to Mississippi and became principal of a military institute. Military schools were numerous in the South. It will be remembered that General W.T. Sherman was engaged in similar work in Louisiana. Stonewall Jackson was professor of military science in Virginia. The South had its full share of cadets in West Point, so that the opening of hostilities found the two sections by no means unequal in the matter of educated officers. Zacharias came north and went out in the Seventh Michigan Infantry, in which he was promoted to captain. He was mortally wounded in the Battle of Antietam. When his body was recovered on the field after the battle, a letter addressed to his father was found clasped in his hand. It read as follows:

"I am wounded, mortally, I think. The fight rages around me. I have done my duty; this is my consolation. I hope to meet you all again. I left not the line until nearly all had fallen and the colors gone. I am weak. My arms are free, but below my chest all is numb. The enemy is trotting over me. The numbness up to my heart. Good-by to all."

ALLAN

The reference, in a previous paragraph, to General Cass, recalls the name of Norval E. Welch, a student of law, who was remarkable for his handsome face and figure. It is related of him that on an occasion when he was in Detroit, he happened to walk past the residence of General Cass, who was then, I believe, one of the United States Senators from Michigan. The latter was so much impressed with the appearance of Welch that he called him back and inquired his name, which was readily given. After a few moments' conversation, Cass asked Welch how he would like to be his private secretary, and receiving a favorable response, tendered him the appointment on the spot. Welch served in that capacity until Cass went into the Cabinet of President Buchanan, when he came to Ann Arbor and took up the study of the law. When the Sixteenth Michigan Infantry was organized, he was commissioned major, and was killed when leaping, sword in hand, over the Confederate breastworks at Peebles's Farm, September 30, 1864. He had in the meantime been promoted to the colonelcy of his regiment.

Morris B. Wells was a graduate of the law department. He

went into the war as an officer of the same regiment with Welch, but was subsequently promoted to be lieutenant colonel of the Twenty-first Michigan Infantry. He was killed at Chickamauga.

No two men could be less alike in appearance than Norval Welch and Morris Wells. One was the embodiment of physical beauty, ruddy with health, overflowing with animal spirits, ready for a frolic, apt with the foils, dumb-bells or boxing gloves, but not particularly a student; the other, tall, rather slender, with an intellectual cast of countenance, frank and manly in his bearing, but somewhat reserved in manner and undemonstrative. Both were conspicuous for their gallantry, but the one impelled by that exuberant physical courage which is distinctive of the leonine type; the other an exemplar of that moral heroism which leads men to brave danger for a principle. They gave everything - even their lives - for their country. The list might be indefinitely extended, but more is not needed to illustrate the spirit of the college boys of 1861-62.

But the students did not all go. Many remained then, only to go later. The prospect of danger, hardship, privation, was the least of the deterrent forces that held them back. To go meant much in most cases. It was to give up cherished plans and ambitions; to abandon their studies and turn aside from the paths that had been marked out for their future lives. Some had just entered that year upon the prescribed course of study; others were half way through; and others still were soon to be graduated. It seemed hard to give it all up. But even these sacrifices were slight compared to those made by older men and heads of families.

And there was no need to depopulate the University at once. The first call filled, those who were left behind began to prepare for whatever might come. The students organized into military companies. *Hardee's Tactics* became the leading textbook. There were three companies or more. These formed a battalion and there was a major to command it. One company was styled "The Tappan Guard," after the venerable President, and it was made up of as fine a body of young men as ever formed in line. Most of them found their way into the Federal army and held good positions.

The captain was Isaac H. Elliott, of Illinois, the athlete, par excellence, of the University, a tall, handsome man and a senior. Tom Weir, a junior, was first lieutenant and the writer second lieutenant. Elliott went to the war as colonel of an Illinois regiment of infantry and was afterwards, for many years, Adjutant General of that state. Weir went out in the Third Michigan Cavalry and became its lieutenant colonel. At the close of the war, he was given a commission as second lieutenant in the Seventh United States Cavalry, Custer's regiment, was brevetted twice for gallantry, and after escaping massacre with his chief at Little Big Horn, died of disease in New York City in 1876.

At the Battle of the Little Bighorn, Captain Tom Weir defied Colonel Benteen's orders to stand down, leading his company on a relief attempt for Custer's beleaguered command. The relief attempt was halted at a hill now known as Weir Point due to pressure from Benteen and Indian resistance.

RECRUITING IN MICHIGAN

ANN ARBOR was not the only town where the fires of patriotism were kept burning. It was one of many "From one learn all." The state was one vast recruiting station. There was scarcely a town of importance which had not a company forming for someone or other of the various regiments that were organizing all through the year. Before the close of the year, aside from the three months men, three regiments of cavalry, eleven regiments of infantry, and five batteries were sent out, all for three years. There was little difficulty in getting recruits to fill these organizations to their maximum standard. No bounties were paid, no draft was resorted to. And yet, the pay for enlisted men was but thirteen dollars a month.

The calls of the President, after the first one for seventy-five thousand, were generally anticipated by the governor, and the troops would be in camp before they were called for, if not before they were needed. The personnel was excellent, and at first great pains were taken to select experienced and competent officers. Alpheus S. Williams, Orlando B. Wilcox, Israel B. Richardson, John C. Robinson, Orlando M. Poe, Thornton F. Brodhead, Gordon Granger, Phillip H. Sheridan and R.H.G. Minty were some of the names that appeared early in the history of Michigan in the war. Under their able leadership, hundreds of young men were instructed in the art of war and taught the principles of tactics, so that they were qualified to take responsible positions in the regiments that were put in the field the following year.

I remember going to see a dress parade of the First Michigan Cavalry at Detroit in August. It was formed on foot, horses not having yet been furnished. It was a fine body of men, and Colonel Thornton F. Brodhead impressed me greatly because of his tall, commanding figure and military bearing. He distinguished himself and was killed at Second Bull Run.

Old City Hall in Detroit, 1865.

Among the other officers was a spare, frail looking man named Town. He was at that time major, and succeeded to the colonelcy after the death of Brodhead. He always sought death on the battlefield, but never found it, and came home to die of consumption after the war was over. He was a modern Chevalier Bayard, and led his regiment at Gettysburg in the grandest cavalry charge of the war. I have no doubt that Meade's right was saved, July 3, 1863, by the superb courage of Charles H. Town and his brave followers. History is beginning to give the cavalry tardy justice for the part it played in that, one of the few great, decisive battles.

One of the most interested spectators of the parade was the venerable statesman and Democratic leader, Lewis Cass. He was then seventy-nine years of age, and few men had occupied a more conspicuous place in State and Nation. He was not without military experience, having been prominent in the frontier War of 1811, and in the War of 1812, he served as an aid to General Harrison. Soon thereafter, he was appointed brigadier general in the United States Army, and was Secretary of War in the Cabinet of President Jackson. He also served as Territorial governor of Michigan under the administrations of Madison, Monroe and John Quincy Adams. The fact of his resignation from the Cabinet of James Buchanan has already been referred to. I confess that I was, for the time being, more interested in that quiet man, standing there under the shadow of a tree, looking on at the parade, than in the tactical movements of the embryonic soldiers.

There was, indeed, much about him to excite the curiosity and inflame the imagination of a youngster only just turned twenty-one. Obtaining a position near where he stood, I studied him closely. He was not an imposing figure, though of large frame, being fat and puffy, with a heavy look about the eyes and a general appearance of senility. He wore a wig. The remarks he made have gone from my memory. They were not of such a character as to leave much of an impression, and consisted mostly of a sort of perfunctory exhortation to the troops to do their duty as patriots.

It was with something of veneration that I looked at this man (standing on the verge of the grave he appeared to be), and yet he outlived many of the young men who stood before him in the bloom of youth. He did not seem to belong to the present so much as to the past. Fifty years before I was born, he had been a living witness of the inauguration of George Washington as first President of the United States. He had watched the growth of the American Union from the time of the adoption of the Constitution. He had been a contemporary of Jefferson, Madison, the Adams, Burr and Hamilton. He had sat in the Cabinets of two different Presidents at widely separated periods. He had represented the government in the diplomatic service abroad, and had served with distinction against the enemies of his country. He had seen the beginning of political parties in the United States and had been a prominent actor through all the changes. He was a youth of twelve when the Reign of Terror in France was in full blast, and thirty-three years of age when Napoleon Bonaparte was on the Island of St. Helena. He had witnessed the downfall of Pitt and the partition of Poland. He was indeed a part of the dead past. His work was done, and it seemed as if a portrait by one of the great masters had stepped down from the canvas to mingle with living persons.

Lewis Cass home. A statue of Cass is one of the two submitted by Michigan to the National Statuary Hall collection in the U.S. Capitol in Washington, D.C. It stands in the National Statuary Hall room. The other is of Gerald Ford, the only U.S. president to come from Michigan.

When the young men from the South who were in the University felt compelled to return to their homes to cast in their lots with their respective states, the students in a body escorted them to their trains and bade them good-by with a sincere wish for good luck to attend them wherever they might go, even though it were into the Confederate military service. The parting was rather with a feeling of melancholy regret that the fates cruelly made our paths diverge, than one of bitterness on account of their belief in the right of states to secede.

Lewis Cass. Cass served as Secretary of State under President James Buchanan. He resigned in 1860 because of Buchanan's failure to mobilize the federal military in the South, an action that in his opinion might have averted the secession of Southern states.

There was a humorous as well as a pathetic side to the war. Soldiers or students, young men were quick to see this. The penchant which boys have to trifle with subjects the most grave gave rise to a funny incident in Ypsilanti (Michigan). There were two rival schools in that town – the State Normal and the Union Seminary. The young men in these two flourishing institutions were never entirely at ease except when playing practical jokes upon each other. Soon after the secession of South Carolina, some of the Seminary boys conceived the idea of compelling the Normal people to show their colors. The first named had put up the Stars and Stripes, a thing that the latter had neglected to do. One morning when the citizens of the town arose and cast their eyes toward the building dedicated to the education and training of teachers, they were astonished to see, flying from the lightning rod on the highest peak of the cupola, a flag of white, whereon was painted a Palmetto tree, beneath the shade of which was represented a rattlesnake in act to strike. How it came there no one could conjecture, but there it was, floating impudently in the breeze, and how to get it down was the question.

I believe that the authorities of the school never learned who it was that performed this daring feat, but it will be violating no confidence at this late day to say that the two heroes of this daring boyish escapade, which was at the time a nine days' wonder, served in the war, one of them in what was known as the "Normal Company," and are now gray-haired veterans, marching serenely down the western slope toward the sunset of their well spent lives.

THE SUMMER OF 1862

THE summer of 1862 was one of the darkest periods of the war. Though more than a year had elapsed since the beginning of hostilities, things were apparently going from bad to worse. There was visible nowhere a single ray of light to illumine the gloom that had settled down upon the land. All the brilliant promise of McClellan's campaign had come to naught, and the splendid Army of Potomac veterans, after having come within sight of the spires of Richmond, was in full retreat to the James. The end seemed farther away than in the beginning.

Grant's successful campaign against Forts Henry and Donelson had been succeeded by a condition of lethargy in all the Western armies. Notwithstanding the successes at Pittsburg Landing and at Corinth, and the death of Albert Sidney Johnston, who had been regarded as the ablest of all the officers of the old army who had taken service with the Confederates, there had been a total absence of decisive results. McClellan had disappointed the hopes of the people; Grant was accused of blundering and of a fondness for drink; the great ability of Sherman was not fully recognized; and the country did not yet suspect that in Sheridan it had another Marlborough. Stonewall Jackson was in full tilt in Virginia, and Robert E. Lee had given evidence that he could easily overmatch any leader who might be pitted against him. With more of hope than of confidence, the eyes of the nation were turned towards Halleck, Buell and Pope.

Richmond, Virginia. Following the attack on Fort Sumter on April 12, 1861, Virginia voted to secede from the United States and join the Confederate States. Soon after, the Confederate government moved its capital from Montgomery, Alabama to Richmond.

It was a dismal outlook. Union commanders were clamoring for more men, and the Union cause was weak because of the lack of confidence which Union generals had in each other. The patriotism of the volunteers, under these most trying and discouraging circumstances, was still the only reliance. Big bounties had not been offered, and the draft had not yet been thought of, much less resorted to. War meetings were being held all over the state, literally in every school house, and recruiting went on vigorously. During the year 1862, Michigan equipped three regiments of cavalry, four batteries, two companies of sharpshooters and fifteen regiments of infantry, which were mustered into the service of the United States.

About the time that the college year closed, President Lincoln issued a call for 300,000 more. This call was dated July 2, 1862, the last previous one having been made on July 25, 1861 - almost a year before. Under this call, Congressman Francis W. Kellogg, of the then Fourth Congressional District of Michigan, came home from Washington with authority to raise two more regiments of cavalry. This authority was direct from Secretary Stanton, with whom for some reason Mr. Kellogg had much influence, and from whom he received favors such as were granted to but few.

He looked like Mr. Stanton. Perhaps that fact may have accounted, in part at least, for the strong bond of friendship between him and the great War Secretary. Under similar authority, he had been instrumental during the year 1861 in putting into the field the Second and Third regiments of Michigan cavalry. They had made an excellent record, and that likewise may have counted to his credit with the War Department. Be that as it may, Mr. Kellogg went at this work with his accustomed vigor, and in a very short space of time, the Sixth and Seventh regiments were ready for muster, though the latter did not leave the state until January, 1863. The Fourth and Fifth regiments had been recruited under a previous call.

To show how little things often change the course of men's lives, an incident of personal experience is here related. The Fifth Michigan Cavalry was recruited under the title of "Copeland's Mounted Riflemen." One of the most picturesque figures in America before the war was John C. Fremont, known as "The Pathfinder," whose "Narrative," in the fifties, was read by boys with the same avidity that they displayed in the perusal of the "*Arabian Nights*." Fremont had a regiment of "Mounted Riflemen" in the Mexican War, though it served in California, and the youthful imagination of those days idealized it into a *corps d'elite*, as it idealized the Mexican War veterans, Marion's men, or the Old Guard of Napoleon Bonaparte.

The name had a certain fascination which entwined it around the memory, and when flaming posters appeared on the walls announcing that Captain Gardner of the village of Muir was raising a company of "Mounted Riflemen" for Copeland's regiment, four young men, myself being one of them, hired a livery team and drove to that modest country four-corners to enlist. The captain handed us a telegram from Detroit saying that the regiment was full and his company could not be accepted. The boys drove back with heavy hearts at the lost opportunity. That is how it happened that I was not a private in the Fifth Michigan Cavalry, instead of a captain in the Sixth when I went out, for in a few days from that time, Mr. Kellogg authorized me to raise a troop, a commission as captain being conditional on my being in camp with a minimum number of men within fifteen days from the date of the appointment.

The conditions were complied with. Two of the other boys became captains in the Sixth Michigan Cavalry; the other went out as sergeant-major of the Twenty-first Michigan Infantry, and arose in good time to be a captain in his regiment.

The government, during the earlier period of the war, was slow to recognize the importance of the cavalry arm of the service. It was expensive to maintain, and the policy of General Scott and his successors was to get along with as small a force of mounted men as possible, and these to be used mostly for escort duty and for orderlies around the various infantry headquarters. There was, consequently, in the cavalry very little of what is known as *"esprit de corps."* In the South, the opposite policy prevailed. At the first Bull Run, the very name of the "Black Horse Cavalry" struck terror into the hearts of the Northern army, though it must be confessed that it was rather moral influence than physical force that the somewhat mythical horsemen exerted. Southern men were accustomed to the saddle, and were as a rule better riders than their Northern brethren. They took naturally to the mounted service, which was wisely fostered and encouraged by the Southern leaders, and under the bold generalship of such riders as Ashby, Stuart, Hampton, Fitzhugh Lee, Rosser, Mosby and others, the cavalry of the Army of Northern Virginia surpassed that of the Army of the Potomac both in numbers and in efficiency.

McClellan says in his book that he often thought he made a mistake in not putting Phil Kearney in command of the cavalry. There is no doubt about it. Kearney had just the right sort of dash. If he had been given a corps of horse with free rein, as Sheridan had it later on, Phil Kearney might have anticipated by at least two years the brilliant achievements of "Cavalry Phil" Sheridan. But the dashing one-armed hero was fated to be killed prematurely, and it was not until 1863 that Pleasanton, Buford, Gregg, Kilpatrick and Custer began to make the Union troopers an important factor in the war; and Sheridan did not take command of the cavalry corps, to handle it as such, until the spring of 1864. Even then, as we shall see later, he had to quarrel with the Commander of the Army in order to compel recognition of its value as a tactical unit upon the field of battle. It was to Hooker, and not to Meade, that credit was due for bringing the cavalry into its proper relation to the work of the Northern army.

McClellan and staff of eight. McClellan is the right of the stump. McClellan was often at odds with President Lincoln over war what Lincoln felt was a tepid war strategy. Prior to relieving McClellan from command, Lincoln met with top generals and directed them to formulate a plan of attack, expressing his exasperation with the remark: "If General McClellan does not want to use the army, I would like to borrow it for a time."

Under the able leadership of such officers as those mentioned, the Federal cavalry took a leading part in the Gettysburg campaign and those which succeeded it, and was able to meet the flower of the South on equal terms and on its own ground. There will be no more honorable page in the history of our country than that on which will be written the record of the cavalry of the armies of the Potomac and of the Shenandoah.

JOINING THE CAVALRY

I FINISHED my sophomore year in June, 1862, and returned to my home full of military spirit and determined to embrace the first favorable opportunity to enter the volunteer service. As second lieutenant of the Tappan Guard, I had acquired a pretty thorough knowledge of Hardee's Tactics and a familiarity with the "school of the soldier" and "school of the company," which proved very useful. Most of the summer was given up to drilling the officers and men in one of the companies of the Twenty-first Michigan Infantry, which was in camp near the town, fitting for the field. The officers were new to the business, without training or experience, as volunteer officers were apt to be, and gladly availed themselves of my help, which was freely given. I was offered a commission as first lieutenant in that regiment, but my ambition was to go in the cavalry, and it was soon to be gratified.

Late in the month of August, my father, coming home from Grand Rapids, met an old friend on the train who told him of Congressman Kellogg's arrival in that place, and what his mission was. I wanted to be a second lieutenant, and told my father that I preferred that to higher rank in the infantry. So the next day, he went down to see the Congressman. His application for my appointment was heartily seconded by a number of influential men in the "Valley City," who knew nothing of me, but did it through their friendship for my father, whom they had known for many years as one of the most energetic and honorable businessmen in the Grand River Valley. From 1848, he had been a familiar figure in lumbering circles, and during that period there had been no year when, from May 1 until snow flew, his fleets of rafts of pine lumber were not running over the dam at Grand Rapids. With the businessmen along the river, his relations had been close and friendly. They were, therefore, not reluctant to do him a favor. Among these, I will mention but two, though there were many others who were equally zealous in the matter.

Michigan was the nation's leading lumber producer between 1869 and 1900. By 1897, over 160 billion board feet was logged from Michigan forests.

Wilder D. Foster and Amos Rathbun were two of the best known men in the metropolis of western Michigan. Mr. Foster was a hardware merchant who had built up a splendid business from small beginnings in the pioneer days. He succeeded Thomas White Ferry in the United States Congress, after Mr. Ferry had been elected to the Senate. Mr. Rathbun, "Uncle Amos" he was called, was a capitalist who had much to do with the development of the gypsum or "plaster" industry in his section of the state. Their influence with Mr. Kellogg was potent, and my father obtained more than he asked for. He came home with a conditional appointment which ran thus:

Headquarters 6th Regt. of Mich. Cavalry
Grand Rapids, Aug. 28, 1862

To Captain James H. Kidd: You are hereby authorized to raise a company of mounted riflemen for this regiment on condition that you raise them within fifteen days from this date, and report with them at the rendezvous in this city.

F.W. KELLOGG, Colonel Commanding

My surprise and gratification can better be imagined than described. To say that I was delighted would be putting it mildly. But the document with the Congressman's signature attached to it was not very much of itself. I was a captain in name only. There was no "company," and would not be unless a minimum of seventy-eight men were recruited, and at the end of fifteen days the appointment would expire by limitation. On the original document which has been carefully preserved appears the following endorsement in Mr. Kellogg's handwriting: "The time is extended for raising this company until Tuesday of next week."

The fifteen days expired on Saturday, and Mr. Kellogg kindly gave us four days extra time to get into camp. It was, however, no easy task to get the requisite number of men in the time allowed, after so many men had been recruited for other regiments. The territory which we could draw upon for volunteers had been very thoroughly canvassed in an effort to fill the quota of the state under Lincoln's last call. But it was less difficult to raise men for cavalry than for infantry, and I was hopeful of succeeding. I soon learned that three others had received appointments for commissions in the same troop - one first, one second and one supernumerary second lieutenant. The same conditions were imposed upon them. Thus, there were four of us whose commissions hinged upon getting a minimum number of men into camp within fifteen days.

Wilder Foster moved to Michigan in 1837, and operated a hardware business in 1845. He was city treasurer, then Mayor of Grand Rapids by 1854. He was a member of the Michigan Senate in 1855 and 1856 and served again as mayor of Grand Rapids in 1865 and 1866.

The man designated for first lieutenant was Edward L. Craw. Some of Craw's friends thought he ought to be the captain, as he was a much older man than myself, though he had no knowledge of tactics and was in every sense a novice in military affairs. In a few days, word came that Mr. Kellogg wanted to see me. He had been told that I was a "beardless boy," and he professed to want men for his captains. My friends advised me not to go - to be too busy recruiting, in fact - and I followed their advice. Had I gone, the colonel would doubtless have persuaded me to change with Craw, since I would have been more than satisfied to take second place, not having too high an opinion of my deserts.

But there was no time to waste, and recruiting was strenuously pushed. Kellogg must have been stuffed pretty full of prejudice, for I never came to town that I did not hear something about it. My friends seemed beset with misgivings. One of them called me into his private office and inquired if I could not manage to raise a beard somehow. I am not sure that he did not suggest a false mustache as a temporary expedient. I told him that it would have to be with a smooth face or not at all. It would be out of the question to make a decent show in a year's time and with careful nursing.

Finally, Uncle Amos Rathbun heard of it and told Kellogg to give himself no concern about "the boy," that he would stand sponsor for him. Uncle Amos, though long ago gathered to his fathers, is alive yet in the memory of hundreds of Union soldiers whom he never failed to help as he had opportunity. And he did not wait for the opportunity to come to him. He sought it. He had a big heart and an open hand, and no man ever had a better friend. As for myself, I recall his name and memory with a heart full of gratitude, for from the moment I entered the service, he was always ready with the needed word of encouragement; prompt with proffers of aid; jealous of my good name; liberal with praise when praise was deserved; appreciative and watchful of my record until the end. If he had faults, they were overshadowed by his kindness of heart and his unaffected virtues. When the record is made up, it will be found with "Uncle Amos" as it was with "Uncle Toby," when he uttered that famous and pardonable oath: "The accusing angel flew to heaven with the oath, blushed as he handed it in. The recording angel, as he wrote it down, dropped a tear upon it and blotted it out forever."

I was the first man to enlist in the embryotic troop and take the oath. The first recruit was Angelo E. Tower, a life-long friend, who entered service as first sergeant and left it as captain, passing through the intermediate grades. His name will receive further mention in the course of this narrative.

The method of obtaining enlistments was to hold war meetings in school houses. The recruiting officer, accompanied by a good speaker, would attend an evening meeting which had been duly advertised. The latter did the talking, the former was ready with blanks to obtain signatures and administer the oath. These meetings were generally well attended, but sometimes it was difficult to induce anybody to volunteer. Once, two of us drove sixteen miles and after a fine, patriotic address of an hour, were about to return without results when one stalwart young man arose and announced his willingness to "jine the cavalry." His name was Solomon Mangus, and he proved to be a most excellent soldier.

On one of my trips, having halted at a wayside inn for lunch, I was accosted by a young man not more than seventeen or eighteen years of age, who said he had enlisted for my troop, and if found worthy, he would be much pleased if he could receive the appointment of eighth corporal. I was amused at the modesty of the request, which was that he be placed on the lowest rung of the ladder of rank. The request did not appear unreasonable, and when the enrollment of troop E, Sixth Michigan Cavalry was completed, he appeared on the list as second corporal. From this rank, he rose by successive steps to that of captain, winning his way by merit alone. For a time, he served on the brigade staff, but whether as corporal, sergeant, lieutenant, captain or staff officer, he acquitted himself with honor and had the confidence of those under whom he served as well as of those whom he commanded. His name was Jacob O. Probasco.

In the western part of the county, our meetings collided with those of Captain Pratt, who had an appointment similar to mine and for the same regiment. Pratt was a big man - a giant almost - full of zeal and enthusiasm. He was a Methodist preacher - a revivalist - and did his own exhorting. He was very fatherly and patronizing, and declared that he would not interfere with my work; that he had plenty of men pledged - more than he needed - and would cheerfully aid in filling my quota in addition to his own.

His promise was taken with a grain of salt, and in the end, I mustered more men than he did, and he had none to spare. Both troops were accepted however, and both of us received our commissions in due time, as the sequel will show.

There was that about "Dominie" Pratt that impressed people with the idea that he would be a great "fighting parson." He was so big, burly and bearded, fierce looking as a dragoon, and with an air of intense earnestness. He was very pious and used to hold prayer meetings in his tent, conducted after the manner of the services at a camp meeting. His confidence in himself, real or assumed, was unlimited. Several of the officers who had seen no service in the field were talking it over one evening in the colonel's tent, and conjecturing how they would feel and act when under fire. Most of them were in anything but a boastful mood, contenting themselves with modestly expressing the belief that when the ordeal came and they were put to the proof, they would stand up to the work and do their duty like officers and gentlemen. Captain Pratt said little, but as we were walking away after the conference had broken up, he placed his arm around my waist in his favorite, affectionate way (he had known me from boyhood) and in his most impressive pulpit manner, said; "Jimmie, (he always addressed me thus) Jimmie, let others do as they may, I want to say to you, that the men who follow me on the field of battle go where death reigneth." As he neared the climax of this dire prediction, he unwound the arm with which he held me to his side and raising it, emphasized his words with a fierce gesture. I confess that I drew back a step, and felt a certain sensation of awe and respect, as I beheld in him the incarnation of courage and carnage.

It may or may not be pertinent to mention that the intrepid captain never led his troop to slaughter; never welcomed the enemy "with bloody hands to hospitable graves." On account of ill health, he was compelled to resign in February, 1863, before the regiment marched from Washington into Virginia. I have always regretted that necessity, because notwithstanding his apparent bravado, the captain was really a brave man, and there was such a fine opportunity in the "Old Dominion" in those days, for one who really hungered for gore to distinguish himself. It would have been a glorious sight to see the gigantic captain, full of the fiery spirit that animated Peter the Hermit when exhorting his followers to the rescue of the holy sepulcher, charging gallantly at the head of his men into the place "where death reigneth." There were several of those places in the southern country.

At the period of the Civil War, the word "company" was applied indiscriminately to cavalry or infantry. The unit of formation was the company. At the present time, there is a distinction. A captain of cavalry commands a "troop." A captain of infantry commands a "company." A troop of cavalry corresponds to a company of infantry. For the sake of convenience and clearness, this classification will henceforth be observed in the course of this narrative.

The troop then, the raising of which has been thus briefly sketched, was ready on Tuesday, September 16, 1862, to begin its career as a military unit in the great army of Union volunteers. It is known in the history of the Civil War as Troop E, Sixth Michigan Cavalry (volunteers). It was originally constituted as follows:

James H. Kidd, captain; Edward L. Craw, first lieutenant; Franklin P. Nichols, second lieutenant; Ambrose L. Soule, supernumerary second lieutenant.

Angelo E. Tower, first sergeant; James L. Manning, quartermaster sergeant; Amos T. Ayers, commissary sergeant; William H. Robinson, William Willett, Schuyler C. Triphagen, Marvin E. Avery, Solon H. Finney, sergeants.

Amos W. Stevens, Jacob O. Probasco, Isaac R. Hart, Benjamin B. Tucker, George I. Henry, David Welch, Marvin A. Filkins, James W. Brown, corporals.

Simon E. Allen, William Almy, Eber Blanchard, Heman S. Brown, Shuman Belding, Lester A. Berry, George Bennett, George Brown, John Cryderman, Edward H. Cook, William B. Clark, James H. Corwin, Eugene C. Croff, William W. Croff, Randall S. Compton, Manley Conkrite, William H. Compton, Seth

Carey, Marion Case, Amaron Decker, Daniel Draper, Rinehart Dikeman, Thomas Dickinson, Orrin W. Daniels, Matthias Easter, Francis N. Friend, Ira Green, James Gray, George I. Goodale, Eli Halladay, Luther Hart, Elias Hogle, George E. Halladay, Robert Hempstead, Edmond R. Hallock, Henry M. Harrison, Warren Hopkins, John J. Hammel, Miles E. Hutchinson, Luther Johnson, Searight C. Koutz, Louis Kepfort, Archibald Lamberton, Martin Lerg, David Minthorn, Solomon Mangus, Andrew J. Miller, Jedediah D. Osborn, Timothy J. Mosher, Gershom W. Mattoon, Moses C. Nestell, George W. Marchant, Edwin Olds, Walter E. Pratt, Albert M. Parker, George W. Rall, Frederick Smith, Jesse Stewart, Josiah R. Stevens, David S. Starks, Orlando V.R. Showerman, David Stowell, James O. Sliter, Jonathan C. Smith, Meverick Smith, Samuel J. Smith, Josiah Thompson, William Toynton, John Tunks, Mortimer Trim, Albert Truax, Oliver L. VanTassel, Byron A. Vosburg, John VanWagoner, Sidney VanWagoner, Erastus J. Wall, Charles Wyman, Harvey C. Wilder, Israel Wall, Lewis H. Yeoman.[2]

The troop that thus started on its career was a typical organization for that time - that is it had the characteristics common to the volunteers of the early period of the Civil War. When mustered into the service, it numbered one hundred and five officers and men. Though for the most part older than the men who went out later, the average age was but twenty-eight years. Nineteen were twenty or under; twenty-nine were thirty or under; eighteen were thirty-one or under. Only nine were over forty. For personnel and patriotism, for fortitude and endurance, they were never excelled. But they were not professional soldiers. At first, they were not soldiers at all. They were farmers, mechanics, merchants, laboring men, students, who enlisted from love of country rather than from love of arms, and were absolutely ignorant of any knowledge of the technical part of a soldier's business.

The militia had been mostly absorbed by the first calls in 1861, and the men of 1862 came from the plow, the shop, the schoolroom, the counting room or the office. With few exceptions, they were not accustomed to the use of arms and had everything to learn. The officers of this particular organization had no advantage over the others in this respect, for save myself, not one of them knew even the rudiments of tactics. Indeed, at the date of muster, there were but three officers in the entire regiment who had seen service. These were Lieutenant Colonel Russell A. Alger, Captain Peter A. Weber and Lieutenant Don G. Lovell.

IN THE REGIMENTAL RENDEZVOUS

IT was a raw, rainy day when we took up the march from the railroad station to the ground whereon had been established the rendezvous for the regiment. It was a motley collection of soldiers, considering the record they were to make during the coming years of active service in the field. All were in citizens' clothes, and equipped with neither uniforms nor arms. Assembled in haste for the journey, there had been no opportunity even to form in line or learn to keep step. No two of them were dressed alike. They were hungry and wet. Few had overcoats, none ponchos or blankets.

Quarters were provided for the night in a vacant store where the men were sheltered from the rain, but had to sleep on the bare floor without cots or comforts of any kind. But, notwithstanding the gloomy conditions that attended this introduction to the volunteer service, they in the main kept up their good spirits, though some were visibly depressed and looked as if they were sorry they had come. In less than a year from that time, they had learned to endure a hundred-fold greater deprivations and hardships with equal minds.

The next morning, breakfast was served in an improvised dining hall on the bank of the river which ran hard by. Then there was another march to camp, the captain reported for duty to the commandant, and a sort of routine of military exercises was entered upon. The officer in command and his adjutant were also new to the business, and haste was made very slowly while they felt their way along. After a few days, the camp was removed to better ground, which was high and dry and overlooked the town. Here, the real work of equipping, organizing and training began.

Winter camp. The camps were set up like small villages complete with crisscrossing lanes called company streets, along with churches and sutler shops.

There were twelve troops, each composed of about one hundred officers and men. The officers were quartered in "wall" tents, but there were not tents enough, so wooden barracks were built for the men. A hospital was established in a house nearby. This was pretty well patronized, at first, the exposure making many men ill. There was a guardhouse also, but not much use for it. A large portion of each day was given up to drill. The rivalry among the captains was spirited, for they had been called together soon after reporting for duty, and informed that they would be given their respective places in line by letter from A to M consecutively, according to proficiency in drill upon a certain date, the two highest places barred, the assignments having been made previously. As the relative rank of these officers depended upon the letter given, it may be imagined that they spared no effort of which they were severally capable. They became immediate students, both in theory and in practice, of Philip St. George Cooke's cavalry tactics wherein the formation in single rank was prescribed.

Soon after going into this camp, uniforms were issued, and horses also. The uniform for the enlisted men at that time consisted of a cavalry jacket, reinforced trousers, forage cap, and boots which came to the knee. Arms, except sabers, were not supplied until after leaving the state. The horses were purchased in Michigan, and great care was taken through a system of thorough inspection to see that they were sound and suitable for the mounted service. In the end, the regiment had a most excellent mount, both the horses and horse equipments being of the best that could be procured.

The horses were sorted according to color, the intention being that each unit should have but one color, as near as practicable. Thus, as I remember it, troop A had bays; B browns; C greys; D blacks; and so on. This arrangement did not last long. A few months' service sufficed to do away with it and horses thereafter were issued indiscriminately. The effect, however, so long as the distinction could be kept up, was fine. It was a grand sight when the twelve hundred horses were in line, formed for parade or drill in single rank, each troop distinguishable from the others by the color of the horses.

When the Fifth Michigan Cavalry was mustered into the United States service at Detroit, there was one supernumerary troop. This was transferred to the Sixth Michigan, then forming in Grand Rapids, and given the letter A without competition. This entitled it to the position on the right flank in battalion formations, and made its commanding officer the senior captain of the regiment. The officers were captain, Henry E. Thompson; first lieutenant, Manning D. Birge; second lieutenant, Stephen H. Ballard; supernumerary second lieutenant, Joel S. Sheldon. Before they left the service, Thompson was lieutenant colonel; Birge, major; Ballard, captain; and Sheldon, regimental commissary.

This troop attracted a great deal of attention from the time of its arrival in camp, for having been organized some two or three months, it was fairly well drilled and disciplined, fully uniformed and the officers were as gay as gaudy dress and feathers could make them. They wore black hats with ostrich plumes and presented a very showy as well as a soldierly appearance. The plumes, like the color arrangement of horses, did not last long. Indeed, few if any of the officers outside of A troop bought them, though they were a part of the uniform prescribed in the books. Two officers who came to the regiment from the Second Michigan Cavalry, and who had had over a year's experience in the field, gave the cue that feathers were not a necessary part of the equipment for real service and served no useful purpose.

One of these two officers I met on the day of my arrival in the temporary camp. It was that wet, drizzly day, when I was sitting in the tent of the commandant awaiting orders. With a brisk step and a military air, a young man of about my own age entered, whose appearance and manner were prepossessing. He looked younger than his years, was not large, but had a well knit, compact frame of medium height. He was alert in look and movement, his face was ruddy with health, his eyes bright and piercing, his head crowned with a thick growth of brown hair cut rather short. He wore a forage cap, a gum coat over his uniform, top boots, and appeared every inch the soldier. He saluted and gave the colonel a hearty greeting, and was introduced to me as Captain Weber.

Peter A. Weber was clerking in a store when the war broke out, and entered service as a corporal in the Third Michigan Infantry. When the Second Michigan Cavalry was organized, he was commissioned battalion adjutant and had been called home to take a captaincy in the Sixth. By reason of his experience, he was given the second place, B. Weber was a rare and natural soldier, the embodiment of courage, and had not death interrupted his career, must have come near the head of the list of cavalry officers. The battle in which he distinguished himself and lost his life will be the theme of a future chapter.

In troop F, commanded by Captain William Hyser, was Second Lieutenant Don G. Lovell, one of the three veteran officers. He went out as corporal in the Third Michigan Infantry, was wounded at Fair Oaks, and again at Trevillian Station while serving in the cavalry. He was one of the bravest of the brave.

Along in September, before the date of muster, I received a letter from a classmate in Ann Arbor asking if there was an opening for him to enlist. I wrote him to come, and soon after joining, he was appointed troop commissary sergeant. At that time, Levant W. Barnhart was but nineteen years of age and a boy of remarkable gifts. He was one of the prize takers in scholarship when he entered the University in 1860, in the class of 1864. His rise in the volunteers was rapid. Passing successively through the grades of first sergeant, second and first lieutenant, he in 1863 was detailed as acting adjutant. While serving in this position, he attracted the notice of General Custer, who secured his appointment by the War Department as assistant adjutant general with the rank of captain. He served on the staff of General Custer until the war closed - succeeding Jacob L. Greene. For one of his age, his record as scholar and soldier was of exceptional brilliancy. He was barely twenty-one when he went on Custer's staff, who was himself not much more than a boy in years. (Custer was but twenty-six when Lee surrendered at Appomattox.)

George Gray, lieutenant colonel commanding, was a lawyer of brilliant parts, a good type of the witty, educated Irishman, a leader at the bar of Western Michigan who had no equal before a jury. He had much reputation as an after dinner speaker, and his polished sentences and keen sallies of wit were greatly enjoyed on occasions where such gifts were in request. Though generally one of the most suave of men, he had an irascible temper at times. The flavor of his wit was tart and sometimes not altogether palatable to those who had to take it. In discipline, he was something of a martinet.

He established a school of instruction in his tent where the officers assembled nightly to recite tactics, and no mercy was shown the luckless one who failed in his lessons. Many a young fellow went away from the "school" smarting under the irony of the impatient colonel. Some of his remarks had a piquant humor, others were characterized by the most biting sarcasm. "Mr. ----," said he one morning when the officers were grouped in front of his tent in response to officers' call, "Mr. ----, have you gloves, sir?" "Yes sir," replied the lieutenant, who had been standing with hands in his trousers pockets. "Well then, you had better put them on and save your pockets." It is needless to say that the young officer thereafter stood in position of the soldier when in presence of his commander.

Nothing was so offensive to Colonel Gray as untidy dress or shabby habiliments on a member of the guard detail. One morning in making his usual inspection, he came upon a soldier who was particularly slovenly. Ordering the man to step out of the ranks, the colonel surveyed him from head to foot, then spurning him with his foot, remarked; "That is a - pretty looking thing for a soldier; go to your quarters, sir."

Once or twice, I felt the sting of his tongue myself, but on the whole he was very kind and courteous, and we managed to get along together very well.

For a time it, was supposed that the colonelcy would go to an army officer, and it may be recalled as an interesting fact that George A. Custer was at that very time a lieutenant on McClellan's staff, and would have jumped at the chance to be colonel of a Michigan cavalry regiment. As has been shown, Philip H. Sheridan, Gordon Granger, O.B. Wilcox, I.B. Richardson and other regulars began their careers as officers in the volunteer service by accepting commissions from Governor Blair. Custer was never a colonel. He was advanced from captain in the Fifth

United States Cavalry to full brigadier general of volunteers, and his first command was four Michigan regiments, constituting what was known as "Custer's Michigan Cavalry Brigade," the only cavalry brigade in the service made up entirely of regiments from a single state.

George Armstrong Custer

A petition was circulated among the officers, asking the governor to appoint Gray colonel. We all signed it, though the feeling was general that it would be better for him to retain the second place and have an officer of the army, or at least one who had seen service, for our commander. The petition was forwarded however, and Gray was commissioned colonel.

Soon thereafter it was announced, greatly to the satisfaction of all concerned, that the vacancy caused by Gray's promotion was to be filled by an officer of experience. Major Russell A. Alger of the Second Michigan Cavalry, who had seen much service in the southwest, was made lieutenant colonel. Major Alger had gone out in 1861 as captain of troop C of the Second Michigan and had earned his majority fighting under Granger and Sheridan. In April, 1861, he was engaged in the lumbering business in Grand Rapids, Michigan, to which place he had removed from Cleveland, Ohio. He had been admitted to the bar in Cleveland, but even at that early day, his tastes and inclinations led him in the direction of business pursuits. He therefore came to Grand River and embarked in lumbering when but just past his majority, and unmarried.

The Panic of 1857 depressed the lumber industry, in common with all other kinds of business, and the young Buckeye met with financial reverses as did nearly everybody in those days, though it is agreed that he showed indications of the dash and self-reliance that were marked features of his subsequent career both in the army and in civil life. Doubtless, had not the war come on, he would have achieved success in his business ventures then as he did afterwards.

Russell Alexander Alger eventually served as Governor and U.S. Senator from Michigan, and also U.S. Secretary of War during the administration of President William McKinley.

When Lieutenant Colonel Alger reported to Colonel Gray for duty, he appeared the ideal soldier. Tall, erect, handsome, he was an expert and graceful horseman. He rode a superb and spirited bay charger which took fences and ditches like a deer. Though not foppish, he was scrupulous to a degree about his dress. His clothes fitted, and not a speck of dust could be found on his person, his horse or his equipments. The details of drill fell largely to him - Colonel Gray attending to the general executive management. As a battalion commander, Colonel Alger had few equals and no superiors. He was always cool and self-poised, and his clear, resonant voice had a peculiar, agreeable quality. Twelve hundred horsemen formed in single rank make a long line, but long as it was, every man could hear distinctly the commands that were given by him.

Weber's voice had the same penetrating and musical quality that made it easy to hear him when he was making no apparent effort to be heard. At that time, it was the custom to give the commands with the voice and not by bugle calls.

Under such competent handling, the regiment soon became a very well drilled organization. The evolutions were at first on foot, then on horseback, and long before the time when it was ready to depart for the front, the officers and men had attained the utmost familiarity with the movements necessary to maneuver a regiment on the field.

On Sundays, it was customary to hold religious services in the camp, and many hundreds of the "beauty and the chivalry" of the town came to see the soldiers and hear the chaplain preach. The regiment would be formed in a hollow square, arms and brasses shining, clothes brushed and boots polished. The chaplain was a good speaker and his sermons were always well worth listening to.

Chaplain Stephen S.N. Greeley was a unique character. Before enlisting, he had been pastor of the leading Congregational church of the city. He was a powerful pulpit orator, a kind hearted, simple minded gentleman of the old school, not at all fitted for the hardships and exposure that he had to undergo while following the fortunes of General Custer's troopers in Virginia. Army life was too much for him to endure, and it was as much as he could do to look after his own physical well being, and the spiritual condition of his flock was apt to be sadly neglected.

Champlain Stephen Greeley

He stayed with the regiment until the end but, in the field he was more like a child than a seasoned soldier, and needed the watchful care of all his friends to keep him from perishing with hunger, fatigue and exposure. I always forgot my own discomforts in commiseration of those of the honest chaplain. When in camp and the weather suitable, I always endeavored to assemble the command for Sunday services, so pleased was he to talk to his "boys." I believe every surviving Sixth Michigan cavalryman has in his heart a warm corner for Chaplain Greeley, who returned to Gilmartin, New Hampshire, the place where he began his ministerial work, and died there many years ago.

While noting in this cursory way the personnel of the regiment, it may be proper to mention the other members of the field and staff. Cavalry regiments were divided into three battalions, each consisting of four troops and commanded by a major. Two troops were denominated a squadron. Thus there were two troops in a squadron, two squadrons in a battalion, three battalions in a regiment. The first major was Thaddeus Foote, a Grand Rapids lawyer. He served with the Sixth about a year and was then promoted to be colonel of the Tenth Michigan Cavalry. Under President Grant, he held the position of pension agent for Western Michigan. Elijah D. Waters commanded the Second Battalion. He resigned for disability and died of consumption in 1866. He did not serve in the field at all. Simeon B. Brown of the Third Battalion was called to the command of the Eleventh Michigan Cavalry in 1863. The Tenth and Eleventh were raised by Congressman Kellogg in that year in the same manner in which he had organized the Second and Third in 1861, and the Sixth and Seventh in 1862.

Speaking of Major Waters recalls how little things sometimes lead on to fortune. After leaving the service, he and his brother started a box factory on the canal in Grand Rapids. In the winter of 1865-66, he took me over to see it. It was a small affair run by water power. The boxes which they manufactured were measures of the old-fashioned kind like the half-bushel and peck measures made of wood fifty years ago. They were of all sizes from a half-bushel down to a quart, and used for "dry measure." Before the top rim was added and the bottom put in, it was customary to pile the cylindrical shells one on top of another in the shop. Looking at these piles one day, Waters saw that three of them, properly hooped, would make a barrel. Why not put hoops on and make them into barrels? No sooner said than done. A patent was secured, a stock company organized and the sequel proved that there were millions in it. The major did not live to enjoy the fruits of his invention, but it made of his brother and partner a millionaire. The latter is today one of the wealthiest men in Michigan - all from that lucky beginning.

The first adjutant of the regiment was Lyman E. Patten, who resigned to become a sutler and was succeeded by Hiram F. Hale, who in turn left the cavalry to become a paymaster. Sutlers were an unnecessary evil; at least, so it seems to me. They were in some cases evil personified. Many of them went into the business solely for the money there was in it, and did not hesitate to trade on the necessities of the boys in blue, so that as a rule there was no love lost, and enlisted men would raid a sutler with as little compunction as the sutler would practice extortion on them.

Commissary tent. Because rations had to be transported long distances, the Commissary Departments relied on foods that could be preserved. The primary meals available to soldiers thus were salted meat and canned goods. Union soldiers also received a hard, cracker-like biscuit that the soldiers dubbed hardtack.

The sutler's tent was too often the army saloon where "S.T. – 1860 - X bitters" and kindred drinks were sold at inflated prices. There were exceptions to the rule however, and Mr. Patten was one of these. The whole sutler business was a mistake. The government should have arranged for an issue, or sale at cost through the commissary and quartermaster departments of such articles as were not regularly furnished and were needed by the officers and men. Sutlers sold a thousand and one things that were not needed and that the men would have been better without. Spirits and tobacco could have been issued as a field or garrison ration, under proper restrictions. This was done at times, but whether a good thing or a bad thing depends altogether upon the point of view. To take up the discussion would be to enter into the controversy as to the army canteen, which is not my purpose.

The medical department of the regiment was in good hands. No officer or enlisted man of the Sixth Michigan ever wanted for kind and sympathetic care when ill or wounded. The position of army surgeon in the field was no sinecure. He had to endure the same privations as the other officers. He was not supposed to be on the fighting line, to be sure, but had to be close at hand to assist in the care of those who were, and oftentimes got into the thickest of it, whether he would or not. To the credit of the profession, be it said, no soldier was ever sick or wounded who did not, unless a prisoner of war, find someone of the green-sashed officers ready to minister to his needs. And it often happened that army surgeons permitted themselves to fall into the enemy's hands rather than to desert those who were under their care and treatment.

The surgeon was Daniel G. Weare, who gave up a lucrative practice to put on the uniform of a major in the medical department of the volunteer army. He was an elderly man with iron grey hair and beard, which became towards the last almost as white as snow. This gave him a venerable look, though this evidence of apparent age was singularly at variance with his fresh countenance, as ruddy as that of youth. He looked like a preacher, though he would swear like a pirate. Indeed, it would almost congeal the blood in one's veins to hear the oaths that came hissing from between the set teeth of that pious looking old gentleman, from whom you would look for an exhortation rather than such expletives as he dealt in. But it was only on suitable provocation that he gave vent to these outbursts, as he was kind of heart, a good friend and a capable physician and surgeon. The assistant was David C. Spaulding, who remained with us but a short time when he was made surgeon of the Tenth Michigan Cavalry - that is to say, in 1863. Weare stayed until the war closed and settled in Fairport, New York, where he died.

Spaulding was surgeon in charge of the regimental hospital in Grand Rapids, and on one occasion came to my aid with some very scientific practice. It happened in this way; it came to my knowledge that a man who had enlisted with one of the lieutenants and mustered in with the troop, was not in the service for the first time; that he had enlisted twice before and then succeeded in getting discharged for disability. The informant intimated that the fellow had no intention of doing duty, would shirk and sham illness and probably get into the hospital, where the chances were he would succeed in imposing on the surgeons and in getting discharged again; that it was pay he was after which he did not propose to earn; least of all would he expose his precious life if by any possibility he could avoid it.

A close watch was put upon the man and sure enough, just before the regiment was to leave the state, he demurred to doing duty, pleading illness as an excuse. I sent him to the hospital, but gave Dr. Spaulding a hint as to the probable nature of the man's illness, and he promised to give his best endeavors to the case. About a week thereafter, the man came back, and whatever might have been his real condition when he went away, he was unmistakably ill. His pale face and weak voice were symptoms that could not be gainsaid. "Well," said I, "have you recovered and are you ready for duty?" "No, I am worse than ever." "Why do you leave the hospital, then?" "My God, captain," whined the man, "they will kill me if I stay there." "But if you are sick, you need treatment." "I cannot enter that place again." "You prefer to perform your duties as a good soldier then?" "I will do anything

72

rather than go there."

He was directed to go about his business and soon thereafter, I inquired about the case. Dr. Spaulding said; "I discovered there was nothing the matter with the man, only that he was playing off, and when he described his alleged symptoms, I began a course of heroic treatment. He was purged, cupped, blistered, given emetics, until life really became a burden and he ran away from the treatment."

This man never went to the regimental hospital again, but he made no end of trouble. He was a chronic shirk. He would not work, and there were not men enough in the regiment to get him into a fight. Soon after the campaign of 1863 opened in Virginia, he was missing, and the next thing heard from him was that he had been discharged from some hospital for disability. He never smelt powder, and years after the war, he was to all appearance an able bodied man. I believe the Sixth was the third regiment which he had gone into in the same way. When he enlisted, the surgeon who examined him pronounced him a sound man, and it was a mystery how he could be physically sound or physically unsound at will, and so as to deceive the medical examiners in either event.

He died long ago, and his widow drew a pension after his death as he did before it, but he never did a day's honest military duty in his life. Peace to his ashes! He may be playing some useful part in the other world, for all that I know. At all events, I am glad that his widow gets a pension, though as a soldier he was never deserving of anything but contempt, for he would desert his comrades when they needed aid and never exposed his precious carcass to danger for his country or for a friend.

That is not an attractive picture which I have drawn. I will paint another, the more pleasing by reason of the contrast which the two present. One day, a party of sixteen men came into camp and applied for enlistment. A condition of the contract under which they were secured for my troop was that one of their number be appointed sergeant. They were to name the man and the choice, made by ballot, fell upon Marvin E. Avery. At first blush, he was not a promising candidate for a non-commissioned office. Somewhat ungainly in figure, awkward in manners and immature in mind and body, he appeared to be; while he seemed neither ambitious to excel nor quick to learn. He certainly did not evince a craving for preferment. In the end, it was found that these were surface indications, and that there were inherent in him a strength of character and a robust manliness that only awaited the opportunity to assert themselves.

He was appointed sergeant, but at first manifested so little aptitude for the work that it was feared he would never become proficient in his duties, or acquire a sufficient familiarity with tactics to drill a squad. No one could have been more willing, obedient or anxious to learn. He was a plodder who worked his way along by sheer force of will and innate self-reliance, and governed in all that he did by a high sense of duty. He never attained first rank as a sergeant while in camp, but in the field, he sprang to the front like a thoroughbred. From the moment when he first scented battle, he was the most valuable man in the troop, from the captain down. In this, I am sure there is no disparagement of the scores of fearless soldiers who followed the guidon of that troop from Gettysburg to Appomattox.

Avery was a hero. In the presence of danger, he knew no fear. The more imminent the peril, the more cool he was. He would grasp the situation as if by intuition, and I often wondered why fate did not make him colonel instead of myself, and honestly believe that he would have filled the position admirably, though he reached no higher rank than that of sergeant. He had, however, made of himself the trusted assistant and adviser of the commanding officer of his regiment, and would have received a commission had he lived but a few days longer. From the day of his enlistment to the day of his death, he was not off duty for a single day; and the command to which he belonged was in no battle when he was not at the front, in the place of greatest risk and responsibility, from the beginning to the end. He was killed by a shell which struck him in the head in the Battle of Trevillian Station, June 12, 1864. A braver or a truer soldier never fell on the field of battle.

Another excellent soldier was Solon H. Finney, who entered service as sergeant. He rose to be second lieutenant and was killed at Beaver Mills, Virginia, April 4, 1865, just five days before Lee surrendered. Finney was a modest, earnest, faithful man, attentive to his duties, not self-seeking, but contented with his lot and ambitious only to do a man's part. It seemed hard for him to go through so near to the end only to be stricken just as the haven of peace was in sight; but his friends have the satisfaction of knowing that Solon Finney never failed to do that which was right, and though he gave his life, it was surrendered cheerfully in the cause of his country and its flag. He was one of those who would have given a hundred lives rather than have his country destroyed - a genuine patriot and a noble man.

With the Washtenaw contingent of troop F came Aaron C. Jewett of Ann Arbor. Jewett was a leading spirit in University circles. His parents were wealthy, he an only son to whom nothing was denied that a doting father could supply. Reared in luxury, he was handsome as a girl and as lovable in disposition. It was current rumor that one of the most amiable young women in the college town - a daughter of one of the professors - was his betrothed. He was graduated with the senior class of that year and immediately enlisted. Notwithstanding his antecedents and his station in life, he performed his humble duties in the ranks without a murmur, thus furnishing one more illustration of the patriotism that animated the best type of young men of that day. Ah, he was a comely soldier, with his round, ruddy face, his fresh complexion, his bright black eyes and curling hair the color of the raven - his uniform brushed and boots polished to the pink of neatness.

These things together, with his modest mien and close attention to his duties, made of him a marked man, and in a short time, regimental headquarters had need of him. He was detailed as clerk, then as acting sergeant-major, and when early in the year 1863 it was announced that Hiram F. Hale was to be appointed army paymaster, Jewett was chosen to succeed him as adjutant, but had not received his commission when death overtook him at Williamsport, Maryland, July 6. There was grief in the Sixth of Michigan on that fateful night when it was known that Aaron Jewett lay within the enemy's lines smitten by a fragment of a shell while faithfully delivering the orders of his colonel to the troops of the regiment as they successively came into line under a heavy fire of artillery.

Williamsport. Fallen Confederate soldiers inside the fence at Hagerstown Road.

Weber and myself, with our men, tried to recover the body but were unable to do so, a force of Confederates having gained possession of the ground. In a week from that time, Weber himself lay cold in death, only five miles distant with a bullet through his brain. That was in Maryland, however, north of the Potomac, and after we had crossed into Virginia, Jewett's father succeeded in finding the body of his son and performed the sad duty of giving it proper sepulture.

All the members of the field and staff of the regiment have been mentioned except Quartermaster Charles H. Patten and Commissary Jacob Chapman. The latter soon resigned. Patten stuck to it until there was no more clothing to issue. He was a good quartermaster, honest, energetic and capable, and that is saying a good deal for him. There has been much uncalled for satirical comment at the expense of the quartermasters. They were really among the most useful of officers - indispensable in fact. The man who handled the transportation for a cavalry command had a position requiring tact, nerve, energy, endurance and ability of a high order. Mr. Patten was such a man. His wagon trains never failed to reach the front with needed supplies when it was possible to get them there. The white canvas of the army wagon was a pleasant sight to the soldier worn out with marching and fighting; and the quartermaster could always count on a cordial welcome when he appeared.

October 11, 1862, the regiment was mustered into the United States service. The mustering officer was General J.R. Smith of the regular army, a veteran of the Mexican War, in which he received a wound in one arm, disabling it. He had a slit in his sleeve tied with ribbons - a way he had, it was thought, of calling attention to his disability, and sort of a standing apology for being back in Michigan while his associates of the army were fighting at the front. It was an amiable and pardonable weakness if such it may be called, and everybody had a liking for the old Mexican War officer.

One of my first acts after reaching the rendezvous had been to call on Colonel Kellogg, who was in his room, up to his eyes in papers and correspondence. He greeted me cordially, congratulated me on my success and assured me that he was my friend, which he proved to be. "Order your uniform at once," said he, "and go to work without delay." The result of this interview was that a tailor took my measure for a suit, and in due time I was arrayed in Union blue, with shining brass buttons, bright yellow facings and the shoulder straps of a captain of cavalry. No boy in his first trousers ever felt happier or prouder.

Wagon train. About twenty-five wagons were needed to supply every thousand men. General Sherman used over five thousand wagons during the Atlanta campaign. His trains would have strung out along sixty miles of road when in line. The order of wagon priority on the narrow roads was ammunition, troops and artillery, followed by quartermaster supplies.

Before the brasses had become tarnished or the trimmings soiled, I took a run to Ann Arbor to say good-bye to the boys. They were glad to see me, and the welcome I had was something to remember. They were like a band of brothers, and showed the same interest as if we had been of one family.

I think the students felt a sort of clannish pride when one of their number enlisted, and thought that the alma mater was doing the correct and patriotic thing in sending her sons into the army. It was plainly to be seen that many of them were holding back unwillingly. Indeed, it was not long until some of them dropped their studies abruptly and followed the example of those who had already gone. Everybody gave me an affectionate Godspeed, and I was surprised at the number of my friends.

THE DEPARTURE FOR WASHINGTON

IT was on a bright moonlight night in December, 1862, that the Sixth Cavalry of Michigan left its rendezvous in Grand Rapids and marched to the station to take the cars for Washington. It was like tearing asunder the ties of years for those whose lines had been cast even for a brief time only in the "Valley City."[3] The hospitality of the people had been unbounded. Many of the officers and men had their homes there. Those who had not took short leaves and made flying visits to their families to say good-bye and arrange their affairs for what might be a final farewell. The scenes of our sojourn for a few months, where we had engaged in daily drills and parades in the pomp and circumstance of mimic warfare, were to know us no longer. The time for rehearsal had passed. We were about to enter upon the real stage of action, and do our part in the mighty tragedy then enacting.

The camp was broken. Tents were struck. Preparations for departure were made. Adieus were said. Horses were sent away in charge of a detail. The quartermaster took possession of the equipments. The regiment was not yet armed, but was to be supplied with all the needed munitions on arrival in the Capital City.

For some reason, it was deemed best to make a night march to the station. No notice of this was given to the citizens. The result was that when we left camp at 2 a.m., the streets were deserted. The town was wrapped in slumber. No sound was heard except the tramp, tramp of the soldiers, and the roar of the river as it plunged over the dam, which only served to intensify the stillness. Through Michigan was a memorable trip. The same scenes with but slight variation were enacted at each station. Officers and men alike were warmed by the hearty and affectionate greetings, the memory of which followed them through all the days, and months, and years of their service.

Michigan Central Railroad Station

On to Detroit, Toledo, Pittsburg, Harrisburg, Baltimore, quickly whirled. Flowers, music, words of cheer everywhere. "God bless you, boys," was the common form of salutation. "Three cheers for the old flag," and "Three cheers for Abe Lincoln," were sentiments offered amidst the wildest enthusiasm, to which the twelve hundred Michigan throats responded with an energy that bespoke their sincerity. Baltimore was reached in the night, and when marching through the streets from one station to the other, the strains of *"John Brown's Body Lies Mouldering in the Ground,"* awoke the echoes in the city that had mobbed a Massachusetts regiment, and through which Abraham Lincoln on the way to his inauguration had to pass in disguise to escape assassination. *"We'll hang Jeff Davis on a sour apple tree,"* was a refrain in which all joined, and there was a heartiness about it that none can understand who did not pass through those troublous times.

But Baltimore was as peaceful as Pittsburg, and no mob gathered to contest the right of Michigan men to invade southern soil. It was quiet. There was no demonstration of any kind. The passage of troops had become a familiar story to the citizens of the Monumental City.

It was the thunder of Burnside's guns at Fredericksburg that welcomed us to the army of the east. The same sun that saw us bivouac beneath the dome of the Capitol shone down upon the Army of the Potomac, lying once again beaten and dispirited on the plains of Falmouth. Burnside had run his course, and "Fighting Joe" Hooker was in command.

THE ARRIVAL IN WASHINGTON

THERE was little about Washington in 1862 to indicate that a great war was raging. The reference in the previous chapter to the "thunder of Burnside's guns" was figurative only. No guns were heard. It was Sunday morning. Church bells pealed out the call for divine worship and streams of well-dressed people were wending their way to the sanctuaries. The presence of uniformed troops in such a scene appeared incongruous, and was the only thing that spoke of war, if we exception the white tents and hospital buildings that abounded on every side.

Mount Pleasant (Harewood) Hospital. Novelist Harriet Beecher Stowe noted, "the city was one great hospital of wounded soldiers; the churches, the public buildings all filled with the maimed, the sick and suffering."

Rest was welcomed after the long jaunt by rail, and the day was given up to it except for the necessary work of drawing and issuing rations. It was historic ground, made doubly so by the events then transpiring. Few realized however, that we actually were engaged in making the history of the most eventful epoch in the career of the Republic, and the chief interest of the place seemed to lie in its associations with the past. The Capitol, with its great unfinished dome, towered above us. The White House, the Treasury building, the Patent office, Arlington, the former home of the Lees, Long Bridge, Pennsylvania Avenue, the Smithsonian Institute, the tree where Sickles killed Key. These and other points of interest were quickly seen or visited.

And the Washington of 1862 was a very different city to the Washington of recent years. Where now are broad avenues of concrete pavement, were then wide streets of mud, through which teams of army mules, hauling heavy wagons, tugged and floundered. A dirty canal full of foul smells traversed the city where now are paved streets and fine buildings. Where then were waste places, now are lovely parks, adorned with statues. Rows of stately trees fringe the avenues and green lawns dot the landscape, where in 1862 was a vast military camp, full of hospitals and squalid in appearance. The man who saw Washington then and returns to it for the first time would be as much astonished as was Aladdin at the creations of his wonderful lamp. Certain salient features remain, but there has been on the whole a magical change.

Camp was pitched on Meridian Hill, well out on Fourteenth Street near Columbia College, then used for a hospital, and preparations were made to spend the winter there. The Fifth Michigan, which had reached Washington before us, was located on "Capitol Hill" at the opposite end of the city. We had a fine campground, stretching from Fourteenth Street through to Seventh, well adapted to drill and parade purposes.

A few days after their arrival in Washington, the officers of the Sixth, under the escort of Congressman Kellogg, went in a body to pay their respects to President Lincoln, several members of the cabinet and the General of the Army. Full dress was the proper caper, they were told, and accordingly they were arrayed in their finest. The uniforms were new, and there is no doubt that they were a gorgeous looking party as they marched up Pennsylvania Avenue wearing shining brasses, bright red sashes, buff gauntlets and sabres glittering in their scabbards. Mr. Kellogg pronounced the "Open Sesame," which caused the doors of the White House to open and secured admission to the presence of the President.

Plans for the building of the new Capitol Dome began in 1857. In 1861, most construction was suspended because of the Civil War, and the Capitol was used briefly as a military barracks, hospital and bakery. In 1862, work on the building was resumed.

After being ushered into the "Blue Parlor," we were kept waiting for some time. Expectancy was on tip-toe, for few if any of the officers had seen Mr. Lincoln. But no introduction was needed when the door opened and the President stood before us. That was to me a memorable moment, for it was the first and last time that I saw Abraham Lincoln. There was no mistaking the tall, gaunt figure, the thin, care-worn face, the slovenly gait, as he entered the room. In appearance, he was almost as unique as his place in history is unexampled. But spare, haggard and bent as he looked, he was yet a strikingly handsome man, for there was on his brow the stamp of greatness. We saw him as in a halo, and looked beyond the plain lineaments and habiliments of the man to the ideal figure of the statesman and president, struggling for the freedom of his country and the unity of his race, whom we all saw in the "Railsplitter" from Illinois; and he seemed, in his absent minded way, to be looking beyond those present to the infinite realm of responsibility and care in which he dwelt.

It is the misfortune of Lincoln that his portraits have not been idealized like those of Julius Caesar, Napoleon Bonaparte and Washington. It remains for some great artist, inspired by the nobility of his subject, to make those homely features so transparent that his reverent and grateful countrymen may look through them and see a presentment of the great soul and beautiful character that irradiated and glorified them in his life, and which will grow brighter and more lovely as the fugitive ages glide away.

The officers were introduced, one by one, and Mr. Lincoln gave each hand a shake as he uttered a perfunctory, but kindly, "How do you do?" and then turned quickly toward the door, as though his mind was still on the work which he had left in order to grant the interview, which must have trenched sadly upon his time.

But he was not to escape so easily, for the Congressman, rising to the occasion, said; "Mr. President, these are the officers of a regiment of cavalry who have just come from my state of Michigan. They are Wolverines, and are on the track of Jeb Stuart, whom they propose to pursue and capture if there is any virtue in a name." "Gentlemen," said the President with a twinkle of the eye, and the first and only indication of humor that he gave, "I can assure you that it would give me much greater pleasure to see Jeb Stuart in captivity than it has given me to see you," and with a bow and smile he vanished.

Although we remained in Washington for about two months, I did not see him again. He never saw Jeb Stuart in captivity, but it was in a fight with the Michigan Cavalry Brigade that the dashing raider was killed. So the remark of the Congressman was not such an idle boast after all.

When the Seventh Michigan arrived, it was put in camp on the Seventh Street side. Colonel J.T. Copeland of the Fifth Michigan was promoted to brigadier general of volunteers and assigned to the command of the three regiments. The brigade was attached to the division of General Silas Casey, all under General S.P. Heintzelman, who was in charge of the Department of Washington, with headquarters in the city. Freeman Norvell succeeded Copeland as colonel of the Fifth. The department extended out into Virginia as far as Fairfax Court House, and there was a cordon of troops entirely around the city.

The prospect was that the brigade would see little, if any fighting, for a time, as it was not to be sent on to the army at Falmouth. The work of drilling and disciplining went on without relaxation throughout the winter months, and when arms were issued, it was found to the delight of all concerned that we were to have repeating rifles.

The muskets or rifles issued to the United States infantry during the Civil War were inferior weapons, and a brigade of Michigan militia of the present period would make short work of a military force of equal numbers so armed. It is one of the strange things about that war that the ordnance department did not anticipate the Austrians, Germans and French in the employment of the firearm loaded at the breech which was so effective in the Franco-Prussian conflict, and if I am not mistaken, in the war between Prussia and Austria in 1866 also. This made of the individual soldier a host in himself. The old muzzle-loader, with its ramrod and dilatory motions, ought to have been obsolete long before Grant left the West to lead the Army of the Potomac from the Wilderness to Appomattox. The Michigan Cavalry Brigade, armed as it was with repeating carbines, was never whipped when it had a chance to use them. In arming the infantry, the government was fifty years behind the times.

Troops in drill. Marching and fighting drill was a part of the daily routine for the Civil War soldier. Soldiers drilled as squads and company formations, each man getting accustomed to orders and formations, how to face properly, dress the line and interact with fellow soldiers. After an hour, the company moved onto regimental level drills and parades.

Possibly the same thing might be said truthfully of the artillery also, though the Union artillerists, notwithstanding the handicap, did such effective work as would have delighted the "Little Corporal" himself.

The Spencer rifle was an invention brought to the notice of the Ordnance Department about that time. Among the numerous charges brought against James G. Blaine was one that he was interested in the manufacture of this arm and in the contract for furnishing it to the government. How much truth there may have been in the assertion I do not know, but if Mr. Blaine was instrumental in bringing about the adoption of the Spencer for the use of the Federal cavalry, he ought to have had a vote of thanks by Congress, for a better gun had never been issued, and if the entire army had been supplied with it, the war could not have lasted ninety days and Mr. Seward would have been a prophet.

The Spencer was a magazine gun carrying eight cartridges, all of which could be discharged without taking the arm from the shoulder. It was loaded at the breech, and the act of throwing out an empty shell replaced it with a fresh cartridge. Against such arms, the old-fashioned muzzle-loaders with which the infantry was equipped were ineffective. The Michigan men were fortunate in being among the very first to receive these repeating rifles, which after the first year in the field were exchanged for the carbine of the same make, a lighter arm and better adapted for the use of cavalry.

THE STAY IN WASHINGTON

THE stay in Washington, though brief, was monotonous. Time hung heavily on our hands. And yet, it was not devoid of incident. There is, perhaps, little of this that is worth recounting of those things, at least that appeared on the surface. Had one been able to reach the penetralia - the inmost recesses - of official and military life, he might have brought away with him reminiscences that would make racy reading. But this privilege was vouchsafed to but few, and they the elect. The logic of war is; learn to obey and ask no questions.

One thing happened which came very near breaking up my troop, and threatened to destroy the regiment itself. It was at that time difficult to get recruits for the regulars. Citizen soldiers preferred the volunteers. But it was considered important to keep the regiments in the regular army recruited up to the minimum at least, and an order was issued from the War Department permitting regular officers to recruit from the ranks of the volunteers. It was a bad order, and as soon as tested was rescinded. I had the misfortune first to experience its effects, and the good fortune to secure its abrogation.

There was in the troop a man who fancied he was slighted when the non-commissioned officers were appointed, and always thereafter nursed his wrath to keep it warm. He was well educated, but of a surly disposition and insubordinate. He was made a corporal, but thought his merits entitled him to something better, and never got over the feeling. Had he gone on and done his duty like General Grant, in the station to which he was assigned, he might have risen much higher. As it was, he never did. This man made the discovery of the War Department order, and soon there was a cabal which was constantly giving out that they were independent of my authority and could shake themselves free at any moment. At first, we did not know what this meant, but it soon leaked out, though they intended to keep it secret. It was ascertained not only that they had the right to go, but that while downtown on passes, eleven men actually had enlisted in the regular army. The recruiting officer had ordered them to report to him on a certain day which they arranged to do, thinking that they would be sent to New York Harbor to garrison forts and escape duty in the field.

When this became known, there was no time to be lost, and Colonel Gray drew up a paper setting forth that if these men were allowed to go, it would be the end of all discipline in his command, and asking that they be ordered to report back for duty. He well understood the art of putting things, and the petition was brief, pointed and convincing. It was addressed to the adjutant general of the army, but had to go through the regular channels, and to save time, he gave me a letter directing that I take it up in person.

In two days, it had been approved by Generals Copeland, Casey and Heintzelman, and there was a delay of one day at that due to a staff officer who acted as a buffer at Heintzelman's headquarters. Proceeding then at once to the adjutant general's office, I was referred to Major Williams,[4] assistant adjutant general, one of the most polished and courteous gentlemen it was ever my fortune to meet. He was most gracious and kind, assured me that the request would be granted at once, and told me to go back and dismiss all further uneasiness about the matter. The next day, the order was rescinded, once and for all. The eleven men were ordered to report back for duty, and the regulars did no more recruiting in the volunteers.

Prior to the Civil War, General Samuel Heintzelman was the president of the Sonora Exploring and Mining Company in Arizona. The town it was headquartered, Cerro Colorado, became famous during the war for the massacre of mine employees by Mexican outlaws, and later for buried treasure.

The men were ignorant of what had been done, and on the morning when they were to leave, they called on me in a body to say good-bye. One of the number, acting as spokesman, assured me that it was on account of no ill will toward captain or troop that they had taken the step. It was done because they believed it would be better for them, and as the act was authorized, begged that I would not think hard of it, at the same time assuring me of their lasting friendship. The speaker doubtless voiced the honest sentiments of all, for it is probable that they themselves had begun to suspect that they were making a mistake. In reply, they were assured that no ill will was harbored, unless it would be in the "harbor" to which they were going, and they were urged to write and let us know how they liked New York Harbor, as we would always feel a warm interest in their welfare.

Then they started, but were halted at the sallyport, and when they exhibited to the officer-of-the-day their passes from the regular army lieutenant, he presented to them the order from the adjutant general. They came back, looking crest fallen enough. Thinking that they had been punished sufficiently, I assured them that if they would do their duty like men, the matter would be forgotten.

It was a good lesson, and from that time on, no officer ever had the honor to command men braver, more faithful or more loyal, than were the regular army contingent of Troop E, Sixth Michigan Cavalry. They never had reason to regret the fate that kept them in the volunteers. Several of them are still living and among my most devoted friends.

At some time during that winter, the Michigan men in Washington had a banquet in one of the rooms or long hallways in the Capitol. It was a fine affair. There were long tables loaded with viands and decorated with flowers. The Michigan Senators - Chandler and J.M. Howard - and the Members of Congress were present, and there was speech making and music. Among those who responded to toasts was Schuyler Colfax, afterwards vice-president, then, I believe, Speaker of the House. Colfax's remarks alone left much of an impression, but I wondered why he was regarded as a great man. He had a pleasant, smiling face and very white teeth, but his speech did not strike one as brilliant in any way.

The singing was led by Doctor Willard Bliss, surgeon-in-charge of Armory Square hospital, located on Fourteenth Street, opposite the then unfinished Washington Monument. Bliss went out as surgeon of the "Old Third,"[5] had already made a place for himself as one of the leading army surgeons, and his hospital was a model of good management. He was at Bull Run with his regiment, and it was said that he sent a telegram from Washington to a relative in Michigan saying; "A great battle fought; Zene (meaning his brother) Zene and I are safe."

Armory Square Hospital had twelve pavilions and overflow tents containing one thousand hospital beds. The wounded were brought to the nearby wharves and then taken to the hospital. It was one of the largest Civil War hospitals in the area and tended to some of the worst injuries due to its close proximity to the wharves, where it could receive the wounded soon after arrival.

The wags were accustomed to figure out what extraordinary time he must have made in order to reach Washington in time to send that telegram. But it was the fashion to guy everybody who was in that battle, unless he was either wounded or taken prisoner. Bliss, as most men are apt to do, "went with the crowd." He remained in Washington after the war, making much money and spending it freely, and achieved notoriety if not fame through his connection with the case of President Garfield after he was shot by the assassin, Guiteau.

The camp on Meridian Hill was a pleasant one, and enlivened at times by the presence of several ladies, among whom were Mrs. Gray, Mrs. Alger and Mrs. Sheldon, wives of the colonel, lieutenant colonel and commissary, respectively. These ladies spent much time in camp, and when the weather was pleasant lived in tents, which always were delightfully homelike and often crowded with visitors. 'Twas but a year or two since Mrs. Alger's soldier-husband led her to the altar as a bride, and they were a handsome couple, not less popular than handsome.

She was a decided favorite in camp, winning the affections of all by her gracious manners and kind heart, as she has done since when presiding over her hospitable home in Detroit or the mansion of the War Secretary in Washington. Mrs. Sheldon, who was a niece of Dr. Willard Bliss, followed her husband to the field and was a ministering angel to many a sick or wounded soldier in hospital and in camp.

One day, a man came to me and wanted to enlist. He said his home was in the State of New York, but he liked the Michigan men and desired to join them. He was a bright looking, active young man, and as the numbers of the troop had been somewhat reduced by sickness and death, he was accepted and mustered in as a private. He remained with us until the morning of the third day at Gettysburg, when about daylight he gathered up a lot of canteens and went, ostensibly, to get them filled. We never saw him again, and many times when thinking of the circumstances, I wondered if he was a Confederate spy. He was a good soldier and did not leave to shirk danger, for he had been under fire and demonstrated his courage. He could hardly have disappeared so completely unless he went into the enemy's lines, and if he did that, must have done it purposely.[6]

There is no doubt that in the early years of the war, the enemy's means of getting information were far superior to ours, and there is still less doubt that not only the army, but Washington, and even the War Department were filled with spies. Probably no Union general ever succeeded in outwitting these Confederate emissaries so completely as did General Sheridan. He told me in Petersburg, after the fall of Richmond, that he had Early's spies at his headquarters in Winchester all through the winter of 1864-65 - they having come to him under the pretense of being deserters - knowing them to be such, but pretending that he did not distrust them, and in the spring, before the grand forward movement, he sent them off on a false scent with wrong information for their chief - Early.

With two of these, in order to keep up the deception, he was obliged to send one genuine Union scout, who was arrested as a spy in Lynchburg, and would have been hung if the sudden closing of hostilities had not suspended sentence. This man's name was M.B. Medes, a trooper of the Sixth Michigan Cavalry, then on detached service as a scout at Sheridan's headquarters, and never since his miraculous escape has he been able to talk about the experiences of that last scout without a fit of nervous prostration. In a letter written to me several years ago, he said: "I don't know why it is, but I can never talk of my adventures and narrow escapes while acting as scout and spy, that I do not break down completely and shake as though I had a hard chill."

FIELD SERVICE IN VIRGINIA

IT was toward the last of February, 1863, that the first order to move came. I had been down to the city, and returning about ten o'clock in the evening, not dreaming of any change from the usual order of things, was surprised to find all bustle and confusion, where a few hours before it had been quiet and serene. The regiment was to march at two o'clock in the morning, and preparations for departure were well under way. Three days' cooked rations and forty rounds of ammunition to the man were to be taken, the sick men and unserviceable horses to remain in camp, and the tents to remain standing as they were until our return. By this it appeared that it was to be a raid or reconnoissance, not a permanent change of station.

Everyone was busy getting ready for the march. Rations were issued, cooked and put in the haversacks; ammunition was distributed and placed in the cartridge boxes; a small bag of oats was strapped to each saddle; horses were fed and the men took a midnight lunch. As for myself, I had the foresight to have a tin cup tied to the cantle of my saddle, and in addition to the cooked meat and hard bread, put into the saddlebags some sugar and a sack of coffee that my good mother had sent from home and which was received only a few days before. It was about as large as a medium sized shot bag, and the coffee was browned and ground ready for use. I also took a supply of matches. These things were of inestimable value during the next few days.

Promptly at the appointed hour, two o'clock a.m., "boots and saddles" and "to horse" were sounded; twelve troops led their horses into line; twelve first sergeants called the roll, to which every man not excused from duty responded; and twelve troop commanders gave the order to mount; when the regiment, responsive to the bugle call "forward," broke into column of fours, moved out into Fourteenth Street and headed for Long Bridge. The night was dark and dismal. The rain began to fall. It was cold and raw, the air surcharged with moisture, chilling one to the marrow. But as the troopers wore gum coats or poncho blankets and top boots, they were measurably sheltered from the storm at the same time that they were exposed to it.

15th Street and Pennsylvania Avenue, Washington DC. A team of oxen haul a 15-inch Rodman gun.

Down through the silent, slumbering city the multitudinous tread of the iron-shod horses awoke strange echoes, while the splashing rain drops and lowering clouds did not serve to raise the spirits. It was an inauspicious beginning of active service, and typical of the many long and weary weeks of wet discomfort that the Sixth of Michigan was destined to experience before the summer solstice had fairly passed. The points of interest, the public buildings, the White House, the massive Greek architecture of the Treasury Building, the Monument, all these as they glided like phantoms through the mist, attracted scarcely a casual glance. Indeed, it is probable that few in that long column took note that these had passed at all, so deeply were they absorbed in the reflections that the time and circumstances produced.

Alexandria, scene of Ellsworth killing. Union Army Colonel Elmer E. Ellsworth had worked at Abraham Lincoln's law office in Illinois. On May 24, 1861, the day after Virginia voters ratified the state decision to secede from the Union, Ellsworth and his troops entered Alexandria, Virginia to assist in the occupation of the city. A Confederate flag large enough to be seen from the White House had been visible in Alexandria for weeks, flown from the roof of an inn, the Marshall House. Ellsworth approached the inn with four troopers. He took down the flag, but as he descended to the main floor, innkeeper James Jackson fired on Ellsworth, killing him instantly. One of Ellsworth's men then fatally shot Jackson. Ellsworth became the first Union casualty of the war.

Thus on to the Long Bridge that spans the great water

highway between the nation's capital and the Old Dominion. The tread of a thousand cavalry horses did not serve to shake its mile of solid superstructure. It seemed a long journey from one end to the other. Above, the scurrying clouds, below, the angry river, all around, the drizzling storm, it was a sorry scene; and a sullen welcome to the soil of Virginia, that was then as often before and afterwards a slippery, sticky mud.

Halting at daylight, the column was reinforced a few miles out by the Fifth Michigan Cavalry. Resuming the march, the two regiments passed through Alexandria, looking with interest, of course, at the spot where the chivalric Ellsworth was shot the year before. What a dilapidated town, its whole face marred and scarred by the ravages of war!

It took till dusk to reach Centerville, and the rain never stopped long enough to catch its breath, but kept at it all day long. Such a first night out as that was! The men slept, or rather stood in the rain all night, for sleep was out of the question. No wood could be procured, so no fires were built and there was no hot coffee. It was a unique experience for cavalrymen, and they had not yet learned how to forage. I wandered around in the rain and finally stumbled upon the quarters of some infantry officers who were stationed near and had a tent and a fire. They kindly permitted me to stay with them until morning. But for this, it seemed to me that I should have perished, though the sequel proved that it was possible to get through a worse night without food or shelter.

In the morning at six o'clock, three more regiments, the Fifth New York, the First Virginia and the Eighteenth Pennsylvania joined, and the force, thus augmented to about two thousand men, pushed on towards Warrenton, Sir Percy Wyndham in command. This officer was an Englishman, an alleged lord. But lord or son of a lord, his capacity as a cavalry officer was not great. He had been entrusted with one or two independent commands and was regarded as a dashing officer. He had no sooner assumed command of our force than he started off at a rapid pace through that part of Virginia that was between Washington and Falmouth - that is, in rear of Hooker's army, and where there was no enemy, unless it might have been small bands of guerrillas.

During the day he charged through the town of Warrenton, and a few Confederate scouts coolly watched the column from the neighboring hills. They were well mounted and evidently did not fear capture. Indeed, no attempt was made to capture them, but away rode Wyndham, as if riding for a wager, or to beat the record of John Gilpin. He seemed bent on killing as many horses as possible, not to mention the men. The fact was the newspapers were in the habit of reporting that Colonel or General so-and-so had made a forced march of so many miles in so many hours, and it is probable that "Sir Percy" was in search of some more of that kind of cheap renown. It was a safe pastime, harmless to the enemy and not dangerous to himself, though hurtful to horse flesh.

That night, we camped beyond Warrenton and had the first taste of picket duty. My troop was sent out about a mile beyond the camp and kept on picket until morning. A line of videttes was posted along the front, and so keenly did the officers feel the responsibility that they made no attempt to sleep, but were in the saddle constantly. It would have been a smart Confederate who could have surprised the Michiganders that night. Every faculty was on the alert. Often we fancied that an enemy was approaching the line; a foe lurked behind every tree and bush; each sound had an ominous meaning, and the videttes were visited at frequent intervals to see if they had discovered anything. In that way, the night passed.

In the morning, everybody was exhausted and to make matters worse, many of the men ran short of provisions. Some of them had neglected to bring the amount ordered; others had been improvident and wasted their rations. So to the discomforts of cold and wet were added the pangs of hunger. The little bag of coffee had proven a precious boon. Whenever the column would halt for a few minutes and it was possible to find anything that would burn, a handful of the coffee was put into a tin cup of water and boiled. It was surprising how quickly this could be done, and the beverage thus brewed was nectar fit for the gods. When the flavor of that coffee, as it tasted on that trip more than forty years ago, is recalled, it is with a smack of the lips. The bare remembrance is more grateful to the palate than is the actual enjoyment of the most delicate product of the culinary art today.

There were times early in the war when spirits were issued to the soldiers as an army ration. Though personally I never took a drop of liquor when on duty during the entire of my army service, yet I am confident that there were times when a reasonable amount of stimulant was a good thing. Indeed, there were times when a man was a fool if he did not take it, assuming that he could get it. Coffee was, however, a very good substitute, and to the credit of the government be it said the coffee issued to the Union troops was almost invariably of excellent quality. They always had it, and plenty of it. Such a solace as it was!

There was nothing like it. On the march, when there was a temporary halt, a thousand fires would quickly blaze alongside the weary column, and a thousand tin cups would soon be steaming with the fragrant and delicious beverage. Veterans could build a fire and make a cup of coffee almost as quickly, and under as discouraging environments, as the traditional Irishman can light his pipe. It seemed to be done by magic, and there was no time and no place where the cup of coffee was not welcome and appreciated.

There is a song, much affected by members of the Grand Army of the Republic. It is styled *"The Army Bean."* I could never quite make out whether it was not intended as a burlesque. There may be enough of sentiment attached to the army bean to entitle it to the honor of being immortalized in song, but to me it was an abomination, less poetic in name and association than the proverbial "sow-belly" bacon, so dear to the heart of the soldier.

Troops in camp. During the campaign season (fair weather), soldiers were in battle only about one day out of thirty. Their remaining days were filled with drilling, punctuated with spells of entertainment such as music, cards and other forms of gambling.

Why does not some poet, filled with the divine afflatus, sing the praise of the army tin cup and its precious contents - the fragrant coffee of the camp, and march, and bivouac? Ambrosial nectar fit for the gods. The everyday and grateful beverage of heroes. Here is a theme for some modern Horace, as inspiring as the fruity and fragrant wine of which his ancient namesake so eloquently sang. I doubt if the red wine of the Horatian odes was more exhilarating to the Roman legionary than the aroma from his tin cup to the soldier of the Union. "Oh, brimming, steaming, fragrant cup! Never failing friend of the volunteer! His solace in fatigue, and his strength in battle. To thee, I sing."

To resume the story at the point at which this digression left it: On the day following the night tour of picket duty, after having ridden from one o'clock in the morning until after eight o'clock in the evening, and the march not yet ended, I became so famished that a piece of raw fat pork was devoured with more relish than ever before I had eaten an orange. Our valiant commander, finding that morning that rations and forage were both exhausted, started for Falmouth, the nearest point at which supplies could be obtained.

Late that Saturday night, we bivouacked with the camp fires of Hooker's army all around. But no forethought had been taken; no rations were drawn or issued; no wood was supplied; and after three days' ride through the rain, many not having had a morsel of food for twenty-four hours, the entire command was forced to lie on the ground in pools of water, in the midst of a drenching rain without food, or fire, or shelter of any kind whatever.

It was dreadful, and the experiences of that night are recalled even now with a shudder. It was like lying down in the middle of a river. There was no place big enough to spread a blanket where there was not a puddle of water, and all the time the rain fell pitilessly in torrents. The solace of hot coffee was denied, for there was no fuel. Food was gone. The minutes were hours. While hunger gnawed at the vitals, a clammy chilliness seized upon one, making him feel as if every vital organ was in a state of congestion. How daylight was longed for, and soon after the first streaks of dawn began to appear, I deserted my watery couch and made straight across the country toward some infantry camps, and actually hugged every fragment of an ember that could be found. After a while, I found some soldiers cooking coffee. One of them was taking a cup off the fire for his breakfast. I asked him for a drink, which he surly refused.

"How much will you take for all there is in the cup?" said I. He did not want to sell it, but when I took out a half dollar and offered it to him, he took it and gave up the coffee, looking on with astonishment while I swallowed it almost boiling hot and without taking breath. This revived me, and soon after, I found a place where a meal consisting of ham, eggs, bread and coffee was served for a big price, and took about a dollar's worth for breakfast.

By eight o'clock, rations and forage were drawn and issued

106

and men and horses were supplied with the much needed food. All of Sunday was spent in Falmouth, and the fresh cavalrymen took a good many observations as to how real soldiers conducted and took care of themselves.

Monday morning, Sir Percy started by the nearest route, via Acquia Creek, Stafford Court House and Fairfax, for Washington, arriving there at eight o'clock Tuesday evening, having been absent just six days, accomplishing nothing. It was a big raid on government horses, ruining a large number. Besides that, it made many men ill. It was a good thing though, after all. The men had learned what campaigning meant, and thereafter knew how to provide themselves for a march, and how important to husband their rations so as to prevent waste at first and make them last as long as possible.

Some idea of the damage done to horses by such raids as that of Sir Percy Wyndham may be gained from the morning reports of officers on the day after the return to camp in Washington. I find that out of eighty horses in my troop, only twenty were fit for duty, part of which had been left in camp and did not accompany the expedition. However, they quickly recuperated, and on the eleventh of March following, we were off into Virginia once more, this time bringing up at Fairfax Court House where we remained a week, encamping by the side of the First Michigan, Fifth New York and several other veteran regiments, from whom by observation and personal contact much information was gained that proved of great value during the following months.

In the meantime, the camps in Washington were broken up and all the regiments were sent across the Potomac. A division of cavalry was organized, consisting of two brigades. Wyndham was sent to Hooker and Julius Stahel, a brigadier general who had been serving in Blenker's division of Sigel's corps, in the Army of the Potomac, was assigned to command of all the cavalry in the Department of Washington, with headquarters at Fairfax Court House.

Fairfax Courthouse. The Fairfax Courthouse was ransacked, its furnishings removed, and the interior generally gutted during the Civil War. The Courthouse was in the years 1863 and 1864 a military outpost and headquarters for the Union Army's supply and communications lines in Northern Virginia.

Stahel was a Hungarian, and it was said had been on the staff of Kossuth in the Hungarian army. He was a "dapper little Dutchman," as everybody called him. His appearance was that of a natty staff officer, and did not fill one's ideal of a major general, or even a brigadier general by brevet. He affected the foreign style of seat on horseback, and it was as good as a show to see him dash along the flank of the column at a rattling pace, rising in his stirrups as he rode. I have always believed that had he remained with the Third Cavalry Division long enough to get into a real charge, like the one at Gettysburg, he would have been glad enough to put aside all those frills and use his thighs to retain his seat in the saddle while he handled his arms.

He took great pride in his messing arrangements, and gave

elegant spreads to invited guests at his headquarters. I was privileged to be present at one of these dinners and must say that he entertained in princely style. His staff were all foreigners, and would have been "dudes," only there were no "dudes" in those days. Dudes were types of the *genus homo* evolved at a later period. They were dandies and no mistake, but in that respect had no advantage over him, for he could vie in style with the best of them. One member of his staff was a Hungarian who answered to the name of Figglemezzy, and only the other day I read a notice of his death recently in New York. Stahel is still living - one of the very few surviving major generals of the Civil War.[7]

It is a pity we did not have a chance to see Stahel in a fight, for I have an idea he was brave, and it takes away in an instant any feeling of prejudice you may have against a man on account of his being fussy in dress when you see him face death or danger without flinching. Fine clothes seem to fit such a man, but upon one who cannot stand fire they become a proper subject for ridicule.

Custer, with flashing eye and flowing hair, charging at the head of his men, was a grand and picturesque figure, the more so by reason of his fantastic uniform which made him a conspicuous mark for the enemy's bullets; but a coward in Custer's uniform would have become the laughing stock of the army. So Stahel might, perhaps, have won his way to confidence had he remained with the cavalry division which afterwards achieved fame under Kilpatrick and Custer, but at the first moment when there was serious work ahead for his command, he was relieved, and another wore the spurs and received the laurels that might have been his.

Leaving Washington at daylight, we went into camp about five miles out, expecting to remain there for a time, but had just time to prepare breakfast when an order came to report to Lieutenant Colonel Alger, who with the four largest troops in the regiment was going off on an independent expedition. That evening we reached Vienna, a little town on the Loudoun Railroad, where we found a small force, including two troops of the First Vermont Cavalry, already on duty. This was our first acquaintance with the Green Mountain Boys, and the friendship thus begun was destined to last as long as there was an enemy in arms against the Union. The First Vermont was sometimes referred to as the "Eighth Michigan," so close were the ties which bound it to the Michigan Brigade. And they always seemed to be rather proud of the designation.

Vienna, cavalry camp. On June 17, 1861, by chance, a Confederate force on a scouting mission heard a Union train approaching Vienna. They set up an ambush along the tracks and with darkness approaching hit the train with two cannon shots. Eight Union soldiers were killed and four wounded. It was very likely the first such combat involving a railroad in any war in world history.

Assuming command of all the forces there, Colonel Alger informed us that General Stahel had information that the place was to be attacked that night and that we were there to defend it. Selecting a strong position on a hill, a camp was started, but no fires were allowed after dark. Vigilance was not relaxed, but no enemy appeared, and on the following day we went on a scout through all the region roundabout without encountering a single armed Confederate. The air was full of rumors. Nobody could tell their origin. Fitzhugh Lee was a few miles away, coming with a big force. Stonewall Jackson had started on another raid, and any moment might see his gray foot cavalry swarming into the vicinity. Such stories were poured into our ears at Vienna, but a couple of days' duty there demonstrated their falsity, and we were hurried back to Fairfax Court House and sent off on a day and night march through the Loudoun Valley to Aldie, Middleburg and Ashby's Gap in the Blue Ridge Mountains.

Two entire regiments, the Fifth Michigan under Colonel Alger and the Sixth under Colonel Gray, went on this expedition, reaching Aldie at midnight in a blinding snow storm. Remaining out in it all night without shelter or fire, the next day we made a gallant charge through Middleburg, finding no enemy there, but a few of Mosby's men who fled at our approach. During the day, some of them were captured, and one man of troop C, Sixth, was killed. It was evident that Lee's army, no portion of it, had begun a movement northward, and the two regiments returned to Fairfax, making a night march while the snow continued to fall and mud and slush made the going as bad as it could be.

At two o'clock in the morning, the column halted and an attempt was made to build camp fires, but the logs and rails were so wet that they would not burn, and all hands stood around in the snow, stamping their feet and swinging their arms in a futile effort to keep warm. The march was resumed at daylight. We were more comfortable when in the saddle, on the march, than during that early morning bivouac. It was possible to sleep when snugly settled in the capacious McClellan saddles, but when dismounted, sleep was out of the question. There was no place to lie down, and to stand in the snow only aggravated the discomfort. But when mounted, the men would pull the capes of their overcoats over their heads, drop their chins upon their breasts and sleep. The horses plodded along and doubtless were asleep too, doing their work as a somnambulist might, walking while they slept.

Soon thereafter, Colonel Alger with five troops (troop B, commanded by Captain Peter A. Weber, having been added to the four that were with him at Vienna) was sent to a place called Camp Meeting Hill, where a camp was established that proved to be a permanent one. At least, we remained there until Hooker's army moved northward. This was a delightful place. The tents were pitched in a grove of large timber on a piece of ground that was high and dry, sloping off in every direction. It was by the side of the pike running south from Vienna, two miles from that place, close to the Leesburg Pike and the Loudoun Railroad. A semi-circular line of pickets was established in front of Washington, the right and left resting on the Potomac, above and below the city respectively.

Our detachment guarded the extreme right of the line. Colonel Gray was five miles to the left, with the remainder of the Sixth and the Fifth still farther away in that direction. About two miles in front of our camp ran the "Difficult" Creek, a small, deep stream with difficult banks that rises somewhere in the Bull Run country and empties into the Potomac near the Great Falls above Washington. A line of videttes was posted along this creek. An enemy could not easily surprise them, as the stream was in their front. Well out toward this line from the main camp, two reserves were established, commanded by captains, and still farther out smaller reserves, under charge of the lieutenants and sergeants. Each troop had a tour of this duty, twenty-four hours on and forty-eight off. The off days were given to reading, writing and exploring the country on horseback.

It was a charming region, not much desolated by the war, being rather out of the beaten track of the armies. Parties of officers often used to take a run across country to Gray's camp, clearing fences and ditches as they went. In these expeditions, Colonel Alger was always the leader, with Captain Weber a close second. On one of these gallopades, he and Weber, who were riding in advance, cleared a stream full of water and about eight or nine feet wide, but when I tried to follow, my horse jumped into instead of across the ditch, the water coming up to the saddle girths. The two lucky horsemen on the other side halted and had a good laugh at my expense, while steed and I were scrambling out the best way we could. My horse was a noble fellow and jumped with all his might when called upon, but lacked judgment and would leap twice as high as was necessary while falling short of making his distance. He rarely failed at a fence, but ditches were a source of dread to horse and man.

The Difficult Creek duty was a sort of romantic episode in our military experience - a delightful green oasis in the dry desert of hard work, exposure, danger and privation. Many pleasant acquaintances were made and time passed merrily. Just across the pike was a spacious farm house, occupied by a family who were staunch Unionists, and who had been made to pay well for their loyalty when the Confederates were in the neighborhood. It was said that Lord Fairfax, the friend of Washington, had at one time lived there. The place had about it an air of generous hospitality that would have become Colonial days. The officers were always welcomed, and it was a favorite resort for them when off duty, partly because the people were Unionists, and partly for the reason that there were several very agreeable young ladies there.

One of these, who lived in Connecticut, was the fiancee of a captain in the First Vermont Cavalry, whose command was stationed there. Another was at home, and it may be surmised that these ladies received the assiduous attentions of half a score, more or less, of the young fellows, who proved themselves thorough cavaliers in gallantry as well as in arms. There was no day when the two ladies might not be seen under the escort of half a dozen cavalrymen, exploring the country on horseback. On all these excursions, Weber, handsome as he was brave, was a leading spirit, and succeeded in captivating the ladies with the charm of his manners, his good looks, his splendid horsemanship and his pleasing address.

It was enough to make one forget the mission that brought him into the South to see him with two or more ladies by his side galloping gaily over the magnificent roads for which that part of Virginia was remarkable. Then there were picnics, lunches, dancing parties and other diversions to fill in the time. Once, one of these parties ventured across the Difficult Creek and rode "between the lines," going as far as Drainesville - eight miles distant - in Mosby's own territory. When the lieutenant colonel commanding learned of this, he reprimanded the officers concerned for what he was pleased to term an act of foolhardiness.

While stationed at this place, one of the young officers was taken ill with fever, and our friends across the way had him brought to the house, where everything that good nursing and kind attention could suggest was done for him. He was reported very ill, and the surgeon said that he was threatened with typhoid fever. A day or two after his removal to the house, I called upon him expecting to find him very low. What was my surprise, on being ushered into a spacious, well furnished apartment, to find him propped up on a bed with a wealth of snowy pillows and an unmistakable look of convalescence, while two good looking ladies sat one on either side of his couch, each holding one of his hands in hers, while he was submitting to the "treatment" with an air of undisguised resignation. It may be noted that this was before the days of "Christian Science." I felt no anxiety about him after that, and returning immediately to camp, wrote to his father stating that if he should hear any rumors that his son was not doing well, to place no reliance upon them, for he was doing very well indeed. This young officer had the good fortune to survive the war, and is still living.

During the sojourn at Difficult Creek, Governor Blair visited the camp. He rode over in the morning on horseback and made an odd looking appearance in his citizen's suit and well worn silk hat. He remained all day, made a speech to the soldiers, and after supper took an ambulance and was escorted by Colonel Alger and myself back to Washington, fourteen miles away. It was a very enjoyable and memorable ride. The war governor was full of anecdote and a good talker, and his companions listened with the liveliest interest to what he had to say about Michigan, her people and her soldiers. He was very solicitous about the welfare of the troops, and impressed one as an able, patriotic man who was doing all he possibly could to hold up the hands of the government, and to provide for the Michigan men in the field. We left him at the National Hotel, and early the next morning returned to our posts of duty.

National Hotel. John Wilkes Booth customarily stayed at the National when in Washington. From there, he studied Lincoln's ways and habits, particularly as a theatre goer. It was at the National where Booth was lodged in the days before he assassinated President Lincoln on April 14, 1865.

About this time, rumors were rife of a projected movement of Lee's army northward. Washington and Alexandria alternated in spasms of fear. Twice, what seemed like well authenticated reports came from the former place that Stuart had passed through our lines. Chain Bridge was torn up, and all the Negroes in Alexandria were out digging rifle pits. Our force was captured repeatedly (without our knowledge) and awful dangers threatened us, according to Washington authority. These and many other equally false reports filled the air. They were probably the result of logical inferences from the actual situation. The time had arrived when active hostilities must soon begin, and what more natural than to suppose that Lee would inaugurate the fray by another invasion of the North?

Among the letters that I wrote to my parents about that time, one or two were preserved, and under date of June 1, 1863, I wrote to my mother a note, the following extract from which will serve to show that there was in our minds a sort of prophetic intuition of what was going to happen. Referring to the false rumors that were not only coming to our ears from these various sources, but even appearing in the Northern papers, I said:

"That Lee will attempt to raid into the North, after the manner of Stonewall Jackson, is possible, perhaps probable, but when he comes we shall hear of it before he wakes up President Lincoln to demand that the keys to the White House be turned over to Jeff Davis. Besides having an efficient and perfect line of pickets, scouts are out daily in our front, so that the idea of the rebel army reaching Washington without our knowledge is preposterous. Lee may make a rapid march through the Shenandoah Valley, and thence into Pennsylvania and Maryland, but nothing would please the Union army more than to have him make the attempt."

Three weeks after the date of that letter, Hooker's army was in motion to head off Lee, who had started to do the very thing thus hinted at, and there was not a soldier in the Federal army of Virginia who did not feel, if he gave the matter any thought, that the Confederate chief had made a fatal mistake, and rejoice at the opportunity to meet him, since meet him we must, outside his entrenchments and the jungles of Virginia. That Stahel's men were willing to do their part was proven by their conduct in the campaign that followed.

During the Civil War, the Chain Bridge was a primary route for the Union Army to access the countryside encampments from Fairfax County. The bridge is the site of the first Union Army Balloon Corps balloon crossing, which took place October 12, 1861, conducted by Professor Thaddeus S. C. Lowe.

Early in June, a thing happened that brought a feeling of gloom into the little camp. Colonel Norvell of the Fifth having resigned, the officers of that regiment united in a petition to the governor to appoint an outsider to the vacancy. Governor Blair selected Lieutenant Colonel Alger. Indeed, that was probably part of his business on the occasion of his recent visit. Colonel Alger was ordered to report immediately for duty with his new command, and left, taking with him the hearty congratulations and good wishes of all his comrades of the Sixth. But their regret at losing him was profound. They did not know how to spare him. It gave him more rank and a larger field of usefulness. Major Thaddeus Foote assumed command of the detachment.

This reference to the Fifth reminds me of Noah H. Ferry and

a night ride in his company, about the time of Colonel Alger's promotion. I had been over to Colonel Gray's camp with some message to him from Colonel Alger, and meeting Major Ferry, who was field officer of the day, he said he was to start that night and inspect the entire picket line of the brigade, about fourteen or fifteen miles long, and invited me to accompany him. He would reach the Difficult outpost in the morning, making an all night ride. I gladly accepted the invitation, both for the ride and to see the country.

Major Ferry, then in his prime, was a strong, vigorous, wholesome looking man with a ruddy complexion and bright eye, a man of excellent habits and correct principles. He told me that night what sacrifices he had made to go into the army. His business had cleared that year $70,000, and with the right sort of management ought to go on prosperously. His leaving it had thrown the entire burden, his work as well as their own, upon the shoulders of his brothers. He had everything to make life desirable - wealth, social position, youth, health - there was nothing to be desired, yet he felt it to be his duty to give it all up to enter the service of his country. He talked very freely of his affairs, and seemed to be weighing in the balances his duty to himself and family. His patriotic feelings gained the mastery however every time, and he talked earnestly of the matter, protesting that our duty to the government in its sore strait ought to outweigh all other considerations.

It was clear that a struggle had been going on in his mind, and that he had resolutely determined to go on and meet his fate, whatever it might be, and when he was killed a few weeks afterwards at Gettysburg, I recalled the conversation of that night and wondered if he had not a presentiment of his coming fate, for he seemed so grave and preoccupied, and profoundly impressed with a sense of the great sacrifice he was making. A soldier neither by profession nor from choice, he wore the uniform of the Union because he could not conscientiously shirk the duty he felt that he owed the government, and relinquished fortune, home, ambition, life itself, for the cause of the Union.

Some time about the middle of June, the picket line was taken up. Major Foote's detachment was ordered to report to Colonel Gray, and Stahel's division was concentrated at Fairfax Court House. The rumors of the movements of armies had become realities. Lee was in motion. The Army of Northern Virginia was trying to steal a march on its great adversary. Long columns of gray were stealthily passing through the Shenandoah Valley to invade the North, and to be on hand to help the farmers of Pennsylvania and Maryland reap their golden harvests.

But the alert Federal commander, gallant "Fighting Joe Hooker," was not caught napping. Lee did not escape from Fredericksburg unobserved. The Army of the Potomac cavalry was sent to guard the passes in the mountains and see to it that Jackson's and Longstreet's maneuvers of the previous summer were not repeated, while six corps of infantry marched leisurely toward the fords of the Potomac, ready to cross into Maryland as soon as it should appear that Lee was actually bent on invasion of Northern soil. Hooker's opportunity had come and he saw it. For Lee to venture into Pennsylvania was to court destruction. All felt that, and it was with elastic step and buoyant spirits that the veterans of Williamsburg and Fair Oaks, of Antietam and Chancellorsville, kept step to the music of the Union as they moved toward the land where the flag was still honored, and where they would be among friends.

All the troops in the Department of Washington were set in motion by Hooker as soon as he arrived where they were. His plan was to concentrate everything in front of Lee, believing that the best way to protect Washington was to destroy the Confederate army. Stahel was ordered to report to General Reynolds, who commanded the left grand division of Hooker's army, and who was to have the post of honor, the advance, and to lose his life while leading the vanguard of the Federal army in the very beginning of the Battle of Gettysburg. Thus it happened that we were at last, part and parcel of that historic army whose fame will last as long as the history of heroic deeds and patriotic endeavor.

Hooker's policy did not coincide with the views of the slow and cautious Halleck, and so the former resigned, thus cutting short a career of extraordinary brilliancy just on the eve of his greatest success. It was a fatal mistake for Hooker. I have always believed that had he remained in command, the Battle of Gettysburg would have been the Appomattox of the Civil War. Such an opportunity as was there presented, he had never had before. Even in the wilderness around Chancellorsville, where his well laid plans miscarried through no fault of his own, he was stopped only by a series of accidents from crushing his formidable adversary. The dense woods prevented the cooperation of the various corps; the audacity of Jackson turned defeat for Lee into temporary victory; and to crown this chapter of accidents, Hooker himself was injured so as to be incapacitated for command at the very moment when quick action was indispensable.

Now the conditions were changed. Jackson, the ablest of all the Confederate generals, was dead, and the Army of the Potomac, greatly reinforced, was to meet the Army of Northern Virginia, materially weakened, where they could have an open field and a fair fight. Every step that Hooker had taken, from the time when he broke camp in Falmouth until he, in a fit of disgust at Halleck's obstinacy, tendered his resignation at Frederick, Maryland, had shown a comprehensive grasp of the situation that inspired the whole army with confidence. The moment that Lee decided to fight the Army of the Potomac on grounds of its own choosing, and to fight an offensive battle, he was foredoomed to defeat, no matter who commanded the Federal army. Hooker possessed the very qualifications that Meade lacked - the same fierce energy that characterized Sheridan - the ability to follow up and take advantage of a beaten enemy. With Hooker in command, Gettysburg would have been Lee's Waterloo.

Sunday, June 21, heavy cannonading in the direction of the passes in the Blue Ridge Mountains proclaimed that the battle was raging. Pleasanton's cavalry had encountered Stuart and Fitzhugh Lee at Middleburg, and a fierce engagement resulted. Our division left Fairfax at an early hour, and all supposed that it would go towards the sound of battle. Not so, however. Stahel, with as fine a body of horse as was ever brought together, marched to Warrenton, thence to Fredericksburg, scouting over the entire intermediate country, encountering no enemy, and all the time the boom of cannon was heard, showing plainly where the enemy was.

We were out three days on this scout, going to Kelly's Ford, Gainesville, Bealton Station, and traversing the ground where Pope's battle of the Second Bull Run was fought, returning by the most direct route to the right of Warrenton. The march was so rapid that the trains were left behind and a good portion of the time we were without forage or food. The horses were fed but once on the trip. Rains had fallen, laying the dust, the weather was charming and it was very enjoyable. One road over which we passed was lined with old cherry trees of the Black Tartarian and Morello varieties, and they were bowing beneath their loads of ripe and luscious fruit with which the men supplied themselves by breaking off the limbs. We passed over much historic ground and were greatly interested in the points where the armies had contended at different times.

IN THE GETTYSBURG CAMPAIGN

AFTER one day of rest from the fatigues of the reconnoissance referred to in the previous chapter, at two o'clock Thursday morning, June 25, the bugles sounded "To Horse," and we bade a final adieu to the places which had known us in that part of the theater of war. The division moved out at daylight. The head of column turned toward Edwards Ferry, on the Potomac River, where Baker fell in 1861. The Sixth was detailed as rear guard. The march was slow, the roads being blocked with wagons, artillery, ambulances, and the other usual impedimenta of a body of troops in actual service, for it was then apparent that the whole army was moving swiftly into Maryland.

At Vienna, the regiment stopped to feed, not being able to move while "waiting for the wagon;" in other words, until all other troops had cleared the way for the rear guard. Vienna was not far from Camp Meeting Hill, so Captain Weber and I obtained permission to ride over and call on our friends in that neighborhood, intending to overtake the regiment at noon. This ride took us two or three miles off the road on which the various commands were marching.

Camp Meeting Hill looked like a deserted village, with no soldiers near and no sign of war. We found our friends rather blue at the thought of being abandoned, and as good-by was said, it was with a feeling that we might never meet again. Weber, gallant as ever, waved his hand to the ladies as he rode away, calling back in a cheery voice that he would come again, "when this cruel war is over." Resuming our journey, a little apprehensive of encountering some of Mosby's men, we were fortunate enough to meet ten troopers of the First Michigan going across the country to join the division. Hurrying on through Dranesville, at a little before noon we overtook the Fifth Michigan Cavalry, from whom we learned that we were up with the advance, and that our own regiment was far in rear.

Selecting a comfortable place, we unsaddled our horses, and lighting our pipes, threw ourselves down on the green grass, and for hours sat waiting while mile after mile of army wagons and artillery passed. Most of the infantry had gone on the day before, but I remember distinctly seeing a portion of the Twelfth Corps en route. I recall especially General A.S. (Pap) Williams and General Geary, both of whom commanded divisions in that corps. At six o'clock in the evening, we went to a farm house and had a supper prepared, but had not had time to pay our respects to it when by the aid of my field glass I saw the advance of the regiment coming. It was the rear guard of a column that was seven hours passing a given point.

It was after dark when the regiment reached the ford at Edwards Ferry. The night was cloudy and there was no moon. The river was nearly, if not quite a mile wide, the water deep and the current strong. The only guide to the proper course was to follow those in advance; but as horse succeeded horse, they were gradually borne farther and farther down the stream, away from the ford and into deeper water. By the time the Sixth reached the river, the water was nearly to the tops of the saddles. Marching thus through the inky darkness, guided mostly by the sound of splashing hoofs in front, there was imminent danger of being swept away and few, except the most reckless, drew a long breath until the distance had been traversed and our steeds were straining up the slippery bank upon the opposite shore.

Safely across the river, the column did not halt for rest or food, but pushed on into Maryland. To add to the discomfort, a drizzling rain set in. The guide lost his way, and it was two o'clock in the morning when the rear guard halted for a brief bivouac in a piece of woods near Poolesville. Wet, weary, hungry and chilled as they were, it was enough to dispirit the bravest men. But there was no murmuring and at daylight, the march was resumed.

That day (26th) we passed the First Army Corps, commanded by the lamented Reynolds, and reached the village of Frederick as the sun was setting. The clouds had cleared away, and a more enchanting vision never met human eye than that which appeared before us as we debouched from the narrow defile up which the road from lower Maryland ran, on the commanding heights that overlooked the valley. The town was in the center of a most charming and fertile country, and around it thousands of acres of golden grain were waving in the sunlight. The rain of the early morning had left in the atmosphere a mellow haze of vapor which reflected the sun's rays in tints that softly blended with the summer colorings of the landscape. An exclamation of surprise ran along the column as each succeeding trooper came in sight of this picture of nature's own painting.

But more pleasing still were the evidences of loyalty which greeted us on every hand as we entered the village. The Stars and Stripes floated above many buildings, while from porch and window, from old and young came manifestations of welcome. The men received us with cheers, the women with smiles and waving of handkerchiefs. That night, we were permitted to go into camp and enjoy a good rest in the midst of plenty and among friends.

On Saturday morning (27th), much refreshed, with horses well fed and groomed and haversacks replenished, the Fifth and Sixth moved on toward Emmittsburg, the Seventh having gone through the Catoctin Valley by another road. The march was through the camps of thousands of infantry just starting in the same direction. Among the distinguished generals who were leading the advance, I remember particularly, Reynolds and Doubleday. During the day, it was a constant succession of fertile fields and leafy woods. Commodious farm houses on every hand and evidences of plenty everywhere, we reveled in the richness and overflowing abundance of the land.

Union troops on the march.

There were oceans of apple-butter and great loaves of snow white bread that "took the cake" over anything that came within the range of my experience. These loaves were baked in brick ovens, out of doors, and some of them looked as big as peck measures. A slice cut from one of them and smeared thick with that delicious apple-butter was a feast fit for gods or men. And then the milk, and the oats for the horses, and everything that hungry man or beast could wish for. Those were fat days and that was a fat country, such as the Iraelitish scouts who went over into the land of Canaan never looked upon or dreamed of.

To be sure, we had to pay for what we had. Especially after we crossed over into Pennsylvania among the frugal Dutch was this the case. But their charges were not exorbitant, and so long as we had a dollar, it was cheerfully parted with for their food. But it seemed a little hard for the Michiganders to be there defending the homes of those opulent farmers, while they, so far from taking up the musket to aid in driving out the army that was invading their soil, were seemingly unwilling to contribute a cent, though I may have misjudged them.

It looked odd too, to see so many able bodied men at home, pursuing their ordinary avocations, with no thought of enlisting, while a hostile army was at their very doors. It looked so to the soldiers who had been serving in Virginia, and who knew that in the South, every man able to bear arms was compelled to do so, and that within the lines of the Confederacy, the cradle and the grave were robbed to fill the ranks. Lee with a hundred thousand men was somewhere in that region, we knew and they knew. We were searching for him, and the time was close at hand when the two armies must come into contact, and oceans of blood would flow before the Confederates could be driven from Northern soil. The government was calling loudly for reinforcements of short time men to serve for the immediate emergency.

Yet, these selfish farmers would drive as sharp a bargain, and figure as closely on the weight and price of an article supplied to the Federal troops, as though they had never heard of war. Indeed, I believe many of them knew little about what was going on. Their world was the little Eden in which they passed their daily lives - the neighborhood in which they lived. They were a happy and bucolic people, contented to exist and accumulate, with no ambition beyond that; and while loyal to the government in the sense that they obeyed its laws and would have scorned to enter into a conspiracy to destroy it, yet they possessed little of that patriotism which inspires men to serve and make sacrifices for their country.

On Sunday morning, June 28, 1863, the two regiments, having passed the night in camp near the Pennsylvania line, resumed the march and passed through the town of Emmittsburg. It was a little place, with scarce more than a thousand inhabitants, but with several churches, an academy, an institute for girls and a little to the northeast, Mount St. Mary's College, a Catholic institution, founded in 1808. Like everything else thereabouts, it had a solid, substantial appearance.

Mount St. Mary's College was founded in 1808 when the preparatory seminary established by the Sulpicians at Pigeon Hill, Pa., was transferred to Emmitsburg. It is the second oldest among the Catholic collegiate institutions in the United States.

So quiet was it, that it seemed like sacrilege to disturb the serenity of that Sabbath day. The sanctuaries stood invitingly in the way, and one could in fancy, almost hear the peal of the organ as the choir chanted *"Gloria in Excelsis,"* glory be to God on high and on earth peace, good will to men, and the voice of the preacher as he read; *"And they shall beat their swords into plowshares and their spears into pruning hooks."*

But our mission was, if possible, to find out what Lee and Longstreet, Ewell and Stuart were doing on that holy day. It required no prophet to predict that it would not be to them a day of rest, but that they would be more than ever active to carry out the schemes that for the Federal army meant great hurt and mischief. Little that was positive was known of Lee's movements, but it was reported that he had pushed on north with his whole army, and was now in dangerous proximity to Harrisburg. His line of march had been to the west of Hooker's and as he was so far north, it was evident that we were making directly for his communications in rear of his army. A tyro in the art of war could see that much of the strategy that was going on. Would Lee allow that and go on to Baltimore, or turn and meet the army that Hooker was massing against him? That was the question.

Gettysburg Cemetery Gate. The Evergreen Cemetery was located on Cemetery Hill at the time of the Battle of Gettysburg in 1863. The brick gatehouse was constructed in 1855. The hill was a key position for the Union Army during the battle.

Taking the Emmittsburg pike, Copeland with the two regiments pushed on to Gettysburg. Thus it was that the Fifth and Sixth Michigan regiments of cavalry had the honor of being the first Union troops to enter the place that was destined so soon to give its name to one of the great battles of history. The road from Emmittsburg to Gettysburg ran between Seminary Ridge on the left and Cemetery Ridge and Round Top on the right. It was a turnpike, and as we marched over it one could not help noticing the strategic importance of the commanding heights on either side. I remember well the impression made on my mind at the time by the rough country off to the right. This was Round Top and Little Round Top where such desperate fighting was done three days later. We passed close to the historic Peach Orchard and over the fish hook shaped Cemetery Hill at the bend; then descended into the town which nestled at the foot of these rocky eminences.

Before we reached the town, it was apparent that something unusual was going on. It was a gala day. The people were out in force, and in their Sunday attire to welcome the troopers in blue. The church bells rang out a joyous peal, and dense masses of beaming faces filled the streets as the narrow column of fours threaded its way through their midst. Lines of men stood on either side with pails of water or apple-butter, and passed a sandwich to each soldier as he passed. At intervals of a few feet were bevies of women and girls who handed up bouquets and wreaths of flowers. By the time the center of the town was reached, every man had a bunch of flowers in his hand or a wreath around his neck. Some even had their horses decorated, and the one who did not get a share was a very modest trooper indeed. The people were overjoyed, and received us with an enthusiasm and a hospitality born of full hearts. They had seen enough of the gray to be anxious to welcome the blue. Their throats grew hoarse with the cheers that they sent up in honor of the coming of the Michigan cavalrymen. The freedom of the city was extended. Every door stood open, or the latch-string hung invitingly out.

Turning to the right, the command went into camp a little outside the town, in a field where the horses were up to their knees in clover, and it made the poor, famished animals fairly laugh. That night, a squadron was sent out about two miles to picket on each diverging road. It was my duty with two troops (E and H) to guard the Cashtown Pike, and a very vivid remembrance is yet retained of the "vigil long" of that July night during which I did not once leave the saddle, dividing the time between the reserve post and the line of videttes. No enemy appeared however, and on Monday (June 29th), the Michigan regiments returned to Emmittsburg, the first cavalry division coming up to take their place in Gettysburg. In this way it came to pass that heroic John Buford, instead of the Fifth and Sixth Michigan, had the honor of meeting the Confederate advance on July 1st.

Before leaving Gettysburg, it was learned that many changes had taken place.[8] Hooker had been succeeded in command of the army by Meade, one of the best and most favorably known of the more prominent generals. It looked like "swapping horses when crossing a stream." Something that touched us more closely, however, was the tidings that Stahel and Copeland had been relieved and that Judson Kilpatrick, colonel of the Second New York (Harris Light) Cavalry had been promoted to brigadier general and assigned to command of the Third Division, by which designation it was thenceforth to be known. He was a West Pointer, had the reputation of being a hard fighter, and was known as "The hero of Middleburg."

Captain Custer of Pleasanton's staff had also received a star, and was to command the Michigan Brigade, to be designated as the Second Brigade, Third Division, Cavalry Corps, Army of the Potomac. Of him we knew but little except that he hailed from Monroe, Michigan, was a graduate of West Point, had served with much credit on the staffs of McClellan and Pleasanton, and that he too was a "fighter." None of us had ever seen either of them. General Copeland turned the two regiments over to Colonel Gray and went away with his staff. I never saw him afterwards.

The Michigan Brigade[9] had been strengthened by adding the First Michigan Cavalry, a veteran regiment that had seen much service in the Shenandoah Valley under Banks, and the Second Bull Run campaign with Pope. It was organized in 1861, and went out under Colonel T.F. Brodhead, a veteran of the Mexican War who was brevetted for gallantry at Contreras and Cherubusco while serving as lieutenant in the Fifteenth United States Infantry.

131

He was mortally wounded August 30, 1862, at Bull Run. His successor was C.H. Town, then colonel of the regiment. He also was severely wounded in the same charge wherein Brodhead lost his life. There had also been added to the brigade Light Battery M, Second United States Artillery, consisting of six rifled pieces and commanded by Lieutenant A.C.M. Pennington.

The Third Division was now ordered to concentrate in the vicinity of Littlestown to head off Stuart, who having made a detour around the rear of the Army of the Potomac, crossed the river below Edwards Ferry on Sunday night, June 28th, and with three brigades under Hampton, Fitzhugh Lee and Chambliss and a train of captured wagons was moving northward, looking for the Army of Northern Virginia, between which and himself was Meade's entire army.

On Monday night, he was in camp between Union Mills and Westminster, on the Emmittsburg and Baltimore Pike, about equidistant from Emmittsburg and Gettysburg. Kilpatrick at Littlestown would be directly on Stuart's path, the direction of the latter's march indicating that he also was making for Littlestown, which place is on a direct line from Union Mills to Gettysburg.

All day of Monday, June 29th, the two regiments (Fifth and Sixth Michigan) were scouting south and east of Gettysburg. Nor did the march end with the day. All night, we were plodding our weary way along, sleeping in the saddle, or when the column in front would halt, every trooper dismounting, and thrusting his arm through the bridle rein, would lie down directly in front of his horse, in the road, and fall into a profound slumber. The horses too would stand with drooping heads, noses almost touching their riders' faces, eyes closed, nodding but otherwise giving no sign, and careful not to step on or injure the motionless figures at their feet. The sound of horses' hoofs moving in front served to arouse the riders when they would successively remount and move on again.

On the morning of June 30th, Kilpatrick's command was

badly scattered. A part of it, including the First and Seventh Michigan and Pennington's battery, was at Abbottstown a few miles north of Hanover; Farnsworth's brigade at Littlestown, seven miles southwest of Hanover. The Fifth and Sixth Michigan arrived at Littlestown at daylight.

Pennington (on right) and battery. Alexander Pennington was born January 8, 1838. He attended West Point, graduated in 1860, and in the Civil War was assigned to the 2nd U.S. Artillery, which would eventually become the first unit to be equipped as "Horse Artillery."

The early morning hours were consumed in scouring the country in all directions, and information soon came in to the effect that Stuart was moving toward Hanover. Farnsworth with the First Brigade left Littlestown for that place at about nine or ten o'clock in the forenoon. The portion of the division that was in the vicinity of Abbottstown was also ordered to Hanover. The Fifth and Sixth Michigan were left, for a time, in Littlestown, troop A of the Sixth under Captain Thompson going on a reconnoissance toward Westminster, and Colonel Alger with the Fifth on a separate road.

The Sixth remained in the town until a citizen came running in, about noon, reporting a large force of the enemy about five miles out toward Hanover. This was Fitzhugh Lee's brigade, and to understand the situation, it will be necessary briefly to describe how Stuart was marching. When he turned off the Baltimore Pike, some seven miles southeast of Littlestown, he had ten miles due north to travel before reaching Hanover. From Littlestown to Hanover is seven miles, the road running northeasterly, making the third side of a right angled triangle. Thus, Stuart had the longer distance to go, and Kilpatrick had no difficulty in reaching Hanover first. Stuart marched with Chambliss leading, Hampton in rear, the trains sandwiched between the two brigades, and Fitzhugh Lee well out on the left flank to protect them.

Farnsworth marched through Hanover, followed by the pack trains of the two regiments that had been left in Littlestown. The head of Stuart's column arrived just in time to strike the rear of Farnsworth, which was thrown into confusion by a charge of the leading Confederate regiment. The pack trains were cut off and captured. Farnsworth however, dashing back from the head of the column, faced the Fifth New York Cavalry to the rear, and by a counter charge, repulsed the North Carolinians and put a stop to Stuart's further progress for that day.

In the meantime, when the citizen came in with the news of Fitzhugh Lee's appearance, "To Horse" was sounded and Colonel Gray led the Sixth Michigan on the Hanover road toward the point indicated. Several citizens with shot guns in their hands were seen going on foot on the flank of the column, trying to keep pace with the cavalry and apparently eager to participate in the expected battle.

When within a mile of Hanover, the regiment turned off into a wheat field, and mounting a crest beyond came upon Fitzhugh Lee's brigade, with a section of artillery in position, which opened upon the head of the regiment (then moving in column of fours) with shell, wounding several men and horses. Lieutenant Potter of troop C had his horse shot under him. Had Gray attacked vigorously, he would have been roughly handled probably, as Fitzhugh Lee was on the field in person with his choice brigade of Virginians. I have always believed however, that a larger force with the same opportunity might have made bad work for Lee.

Hanover Junction Railway Station. On November 18-19, 1863, President Abraham Lincoln traveled through Hanover Junction on to and from dedication ceremonies for the Gettysburg National Cemetery.

Colonel Gray, seeing that the force in front of him were preparing to charge, and aware that one raw regiment would be no match for a brigade of veteran troops, made a detour to the left, and sought by a rapid movement to unite with the command in Hanover, Major Weber with troops B and F being entrusted with the important duty of holding the enemy in check while the others effected their retreat. Right gallantly was this duty performed. Three charges upon the little band were as often repulsed by the heroic Weber, and with such determination did he hold to the work that he was cut off and did not succeed in rejoining the regiment until about three o'clock the next morning. Colonel Alger with the Fifth and troop A of the Sixth under Captain H.E. Thompson, also had a smart encounter with the same force, holding their own against much superior numbers by the use of the Spencer repeating rifles with which they were armed.

By noon or soon after, the entire division united in the village of Hanover. The First, Fifth, Sixth and Seventh Michigan regiments and Pennington's battery were all on the ground near the railroad station. The Confederate line of battle could be distinctly seen on the hills to the south of the town. The command to dismount to fight on foot was given. The number one, two and three men dismounted and formed in line to the right facing the enemy. The number four men remained with the horses, which were taken away a short distance to the rear.

It was here that the brigade first saw Custer. As the men of the Sixth, armed with their Spencer rifles, were deploying forward across the railroad into a wheat field beyond, I heard a voice new to me, directly in rear of the portion of the line where I was, giving directions for the movement in clear, resonant tones and in a calm, confident manner, at once resolute and reassuring. Looking back to see whence it came, my eyes were instantly riveted upon a figure only a few feet distant whose appearance amazed if it did not for the moment amuse me. It was he who was giving the orders. At first, I thought he might be a staff officer, conveying the commands of his chief. But it was at once apparent that he was giving orders, not delivering them, and that he was in command of the line.

Looking at him closely, this is what I saw: An officer superbly mounted who sat his charger as if to the manor born. Tall, lithe, active, muscular, straight as an Indian and as quick in his movements, he had the fair complexion of a school girl. He was clad in a suit of black velvet, elaborately trimmed with gold lace which ran down the outer seams of his trousers and almost covered the sleeves of his cavalry jacket. The wide collar of a blue navy shirt was turned down over the collar of his velvet jacket, and a necktie of brilliant crimson was tied in a graceful knot at the throat, the long ends falling carelessly in front. The double rows of buttons on his breast were arranged in groups of twos, indicating the rank of brigadier general. A soft, black hat with wide brim adorned with a gilt cord, and rosette encircling a silver star, was worn turned down on one side giving him a rakish air. His golden hair fell in graceful luxuriance nearly or quite to his shoulders, and his upper lip was garnished with a blonde mustache. A sword and belt, gilt spurs and top boots completed his unique outfit.

George A. Custer, 1863

A keen eye would have been slow to detect in that rider with the flowing locks and gaudy tie, in his dress of velvet and of gold, the master spirit that he proved to be. That garb, fantastic as at first sight it appeared to be, was to be the distinguishing mark which during all the remaining years of that war, like the white plume of Henry of Navarre, was to show us where, in the thickest of the fight, we were to seek our leader - for where danger was, where swords were to cross, where Greek met Greek, there was he, always.

Brave but not reckless; self-confident yet modest; ambitious but regulating his conduct at all times by a high sense of honor and duty; eager for laurels but scorning to wear them unworthily; ready and willing to act but regardful of human life; quick in emergencies, cool and self-possessed, his courage was of the highest moral type, his perceptions were intuitions. Showy like Murat, fiery like Farnsworth, yet calm and self-reliant like Sheridan, he was the most brilliant and successful cavalry officer of his time. Such a man had appeared upon the scene, and soon we learned to utter with pride the name of Custer.

George A. Custer was, as all agree, the most picturesque figure of the Civil War. Yet his ability and services were never rightly judged by the American people. It is doubtful if more than one of his superior officers - if we except McClellan, who knew him only as a staff subaltern - estimated him at his true value. Sheridan knew Custer for what he was. So did the Michigan Brigade and the Third Cavalry Division. But, except by these, he was regarded as a brave, dashing but reckless officer who needed a guiding hand. Among regular army officers as a class, he cannot be said to have been a favorite. The meteoric rapidity of his rise to the zenith of his fame and success, when so many of the youngsters of his years were moving in the comparative obscurity of their own orbits, irritated them. Stars of the first magnitude did not appear often in the galaxy of military heroes. Custer was one of the few.

The popular idea of Custer is a misconception. He was not a reckless commander. He was not regardless of human life. No man could have been more careful of the comfort and lives of his men. His heart was tender as that of a woman. He was kind to his subordinates, tolerant of their weaknesses, always ready to help and encourage them. He was brave as a lion, fought as few men fought, but it was from no love of it. Fighting was his business; and he knew that by that means alone could peace be conquered. He was brave, alert, untiring, a hero in battle, relentless in the pursuit of a beaten enemy, stubborn and full of resources on the retreat. His tragic death at the Little Big Horn crowned his career with a tragic interest that will not wane while history or tradition endure. Hundreds of brave men shed tears when they heard of it - men who had served under and learned to love him in the trying times of Civil War.

I have always believed that some of the real facts of the Battle of the Little Big Horn were unknown. Probably the true version of the massacre will remain a sealed book until the dead are called upon to give up their secrets, though there are those who profess to believe that one man at least is still living who knows the real story and that some day he will tell it.

Certain it is that Custer never would have rushed deliberately on destruction. If, for any reason, he had desired to end his own life, and that is inconceivable, he would not have involved his friends and those whose lives had been entrusted to his care in the final and terrible catastrophe. He was not a reckless commander or one who would plunge into battle with his eyes shut. He was cautious and wary, accustomed to reconnoiter carefully and measure the strength of an enemy as accurately as possible before attacking.

More than once, the Michigan Brigade was saved from disaster by Custer's caution. This may seem to many a novel - to some an erroneous estimate of Custer's characteristics as a military man. But it is a true one. It is an opinion formed by one who had good opportunity to judge of him correctly. In one sense only is it a prejudiced view. It is the judgment of a friend and a loyal one; it is not that of an enemy or a rival. As such, it is appreciative and it is just.

Under his skillful hand, the four regiments were soon welded into a coherent unit, acting so like one man that the history of one is oftentimes apt to be the history of the other, and it is difficult to draw the line where the credit that is due to one leaves off and that which should be given to another begins.

The result of the day at Hanover was that Stuart was driven still farther away from a junction with Lee. He was obliged to turn to the east, making a wide detour by the way of Jefferson and Dover Kilpatrick, meanwhile, maintaining his threatening attitude on the inside of the circle which the redoubtable Confederate was traversing, and forcing the latter to swing clear around to the north as far as Carlisle, where he received the first reliable information as to the whereabouts of Lee. It was the evening of July 2nd when he finally reached the main army. The battle then had been going on for two days, and the issue was still in doubt. During that day (2), both Stuart and Kilpatrick were hastening to rejoin their respective armies, it having been decided that the great battle would be fought out around Gettysburg. Gregg's division had been guarding the right flank of Meade's army, but at nightfall it was withdrawn to a position on the Baltimore Pike near the reserve artillery.

Kilpatrick reached the inside of the Union lines in the vicinity of Gettysburg late in the afternoon, at about the same hour that Hampton, with Stuart's leading brigade, arrived at Hunterstown a few miles northeast of Gettysburg. It was about five o'clock in the afternoon when the Third Division, moving in column of fours, was halted temporarily, awaiting orders where to go in and listening to the artillery firing close in front, when a staff officer rode rapidly along the column, crying out; "Little Mac is in command and we are whipping them." It was a futile attempt to evoke enthusiasm and conjure victory with the magic of McClellan's name. There was scarcely a faint attempt to cheer. There was no longer any potency in a name.

Soon thereafter, receiving orders to move out on the road to Abbottstown, Kilpatrick started in that direction, Custer's brigade leading with the Sixth Michigan in advance. When nearing the village of Hunterstown on a road flanked by fences, the advance encountered a heavy force of Confederate cavalry. A mounted line was formed across the road, while there were dismounted skirmishers behind the fences on either side. The leading squadron of the Sixth, led by Captain H.E. Thompson, boldly charged down the road, and at the same time, three troops were dismounted and deployed on the ridge to the right, Pennington's battery going into position in their rear.

Hunterstown, PA. While the major fighting was taking place just four miles southwest at Gettysburg, two Union brigades attempted to probe the left flank and rear of the Confederate Army of Northern Virginia by using the road that connected Hunterstown to Gettysburg. A Confederate cavalry brigade arrived on the road near the village and obstructed the Union advance. The skirmish lasted until after dark, when the Union cavalry withdrew southeast.

The mounted charge was a most gallant one, but Thompson, encountering an overwhelmingly superior force in front and exposed to a galling fire on both flanks as he charged past the Confederates behind the fences, was driven back, but not before he himself had been severely wounded, while his first lieutenant, S.H. Ballard, had his horse shot under him and was left behind a prisoner. As Thompson's squadron was retiring, the enemy attempted a charge in pursuit, but the dismounted men on the right of the road kept up such a fusillade with their Spencer carbines, aided by the rapid discharges from Pennington's battery, that he was driven back in great confusion.

General Kilpatrick, speaking in his official report of this engagement, says: "I was attacked by Stuart, Hampton and Fitzhugh Lee near Hunterstown. After a spirited affair of nearly two hours, the enemy was driven from this point with great loss. The Second Brigade fought most handsomely. It lost in killed and wounded and missing, 32. The conduct of the Sixth Michigan Cavalry and Pennington's battery is deserving of the highest praise."

On the other hand, General Hampton states that he received information of Kilpatrick's advance upon Hunterstown and was directed by Stuart to go and meet it. He says: "After some skirmishing, the enemy attempted a charge, which was met in front by the Cobb Legion, and on either flank by the Phillips Legion and the Second South Carolina Cavalry."

The position at Hunterstown was held until near midnight when Kilpatrick received orders to move to Two Taverns on the Baltimore turnpike, about five miles southeast of Gettysburg and some three miles due south from the Rummel farm, on the Hanover road east of Gettysburg, where the great cavalry fight between Gregg and Stuart was to take place on the next day. It was three o'clock in the morning (Kilpatrick says daylight) when Custer's brigade went into bivouac at Two Taverns.

The Second Cavalry Division, commanded by General D. McM. Gregg, as has been seen, held the position on the Rummel farm on the 2nd, but was withdrawn in the evening to the Baltimore Pike "to be available for whatever duty they might be called upon to perform on the morrow." On the morning of the third, Gregg was ordered to resume his position of the day before, but states in his report that the First and Third Brigades (McIntosh and Irvin Gregg) were posted on the right of the infantry, about three-fourths of a mile nearer the Baltimore and Gettysburg pike, because he learned that the Second Brigade (Custer's) of the Third Division was occupying his position of the day before.

General Kilpatrick, in his report says: "At 11 p.m. (July 2nd) received orders to move (from Hunterstown) to Two Taverns, which point we reached at daylight. At 8 a.m. (July 3rd) received orders from headquarters cavalry corps to move to the left of our line and attack the enemy's right and rear with my whole command and the Reserve Brigade. By some mistake, General Custer's brigade was ordered to report to General Gregg and he (Custer) did not rejoin me during the day."

General Custer, in his report, gives the following, which is without doubt, the true explanation of the "mistake." He says: "At an early hour on the morning of the third, I received an order through a staff officer of the brigadier general commanding the division (Kilpatrick), to move at once my command and follow the First Brigade (Farnsworth) on the road leading from Two Taverns to Gettysburg. Agreeably to the above instructions, my column was formed and moved out on the road designated, when a staff officer of Brigadier General Gregg, commanding the Second Division, ordered me to take my command and place it in position on the pike leading from York[10] (Hanover) to Gettysburg, which position formed the extreme right of our line of battle on that day."

Thus it is made plain that there was no mistake about it. It was Gregg's prescience. He saw the risk of attempting to guard the right flank with only the two decimated brigades of his own division. Seeing with him was to act. He took the responsibility to intercept Kilpatrick's rear and largest brigade, turn it off the Baltimore pike to the right instead of allowing it to go to the left as it had been ordered to do, and thus doubtless, a serious disaster was averted. It makes one tremble to think what might have been, of what inevitably must have happened, had Gregg, with only the two little brigades of McIntosh and Irvin Gregg and Randol's battery, tried to cope single handed with the four brigades and three batteries, comprising the very flower of the Confederate cavalry and artillery, which those brave knights - Stuart, Hampton and Fitzhugh Lee - were marshaling in person on Cress's Ridge.

If Custer's presence on the field was, as often has been said, providential, it is General D. McM. Gregg to whom, under Providence, the credit for bringing him there was due. Gregg was a great and a modest soldier and it will be proper, before entering upon a description of the battle in which he played so prominent a part, to pause a moment and pay to him the merited tribute of our admiration. In the light of all the official reports, put together link by link so as to make one connected chain of evidence, we can see that the engagement which he fought on the right at Gettysburg, on July 3rd, 1863, was from first to last a well planned battle in which the different commands were maneuvered with the same sagacity displayed by a skillful chess player in moving the pawns upon a chessboard; in which every detail was the fruit of the brain of one man who, from the time when he turned Custer to the northward until he sent the First Michigan thundering against the brigades of Hampton and Fitzhugh Lee, made not a single false move; who was distinguished not less for his intuitive foresight than for his quick perceptions at critical moments. That man was General David McMutrie Gregg.

This conclusion has been reached by a mind not – certainly not - predisposed in that direction, after a careful study and review of all the information within reach bearing upon that eventful day. If at Gettysburg, the Michigan Cavalry Brigade won honors that will not perish, it was to Gregg that it owed the opportunity, and his guiding hand it was that made its blows effective. It will be seen how later in the day he again boldly took responsibility at a critical moment and held Custer to his work on the right, even after the latter had been ordered by higher authority than himself (Gregg) to rejoin Kilpatrick and after Custer had begun the movement.

General Gregg (seated) and staff. Recognized are Captain Albert M. Harper, Captain Treishel, Colonel C. Taylor and Captain Henry C. Weir.

Now, having admitted if not demonstrated that Gregg did the planning, it will be shown how gallantly Custer and his Michigan Brigade did their part of the fighting. Up to a certain point, it will be best to let General Custer tell his own story: "Upon arriving at the point designated, I immediately placed my command in a position facing toward Gettysburg. At the same time, I caused reconnoissances to be made on my front, right and rear, but failed to discover any considerable force of the enemy. Everything remained quiet until 10 a.m., when the enemy appeared on my right flank and opened upon me with a battery of six guns."

"Leaving two guns and a regiment to hold my first position and cover the road leading to Gettysburg, I shifted the remaining portion of my command forming a new line of battle at right angles with my former position. The enemy had obtained correct range of my new position, and was pouring solid shot and shell into my command with great accuracy. Placing two sections of Battery M, Second Regular Artillery, in position, I ordered them to silence the enemy's battery, which order, notwithstanding the superiority of the enemy's position, was done in a very short space of time."

"My line as it then existed was shaped like the letter "L." The shorter branch, supported by one section of Battery M (Clark's), supported by four squadrons of the Sixth Michigan Cavalry, faced toward Gettysburg, covering the pike; the long branch, composed of the two remaining sections of battery M, supported by a portion of the Sixth Michigan Cavalry on the left, and the First Michigan Cavalry on the right - with the Seventh Michigan Cavalry still further to the right and in advance - was held in readiness to repel any attack on the Oxford (Low Dutch) Road.[11]"

"The Fifth Michigan was dismounted and ordered to take position in front of my center and left. The First Michigan was held in column of squadrons to observe the movements of the enemy. I ordered fifty men to be sent one mile and a half on the Oxford (Low Dutch) Road, and a detachment of equal size on the York (Hanover) Road, both detachments being under the command of the gallant Major Weber (of the Sixth) who from time to time kept me so well informed of the movements of the enemy, that I was enabled to make my dispositions with complete success."

General Custer says further that at twelve o'clock, he received an order directing him, on being relieved by a brigade of the Second Division, to move to the left and form a junction with Kilpatrick; that on the arrival of Colonel McIntosh's brigade he prepared to execute the order; but, to quote his own language: "Before I had left my position, Brigadier General Gregg, commanding the Second Division, arrived with his entire command. Learning the true condition of affairs, and rightly conjecturing the enemy was making his dispositions for vigorously attacking our position, Brigadier General Gregg ordered me to remain in the position I then occupied."

So much space has been given to these quotations because

they cover a controverted point. It has been claimed, and General Gregg seems to countenance that view, that Custer was withdrawn and that McIntosh, who was put in his place, opened the fight, after which Gregg brought Custer back to reinforce McIntosh. So far from this being true, it is quite the reverse of the truth. Custer did not leave his position. The battle opened before the proposed change had taken place, and McIntosh was hurried in on the right of Custer. The latter was reluctant to leave his post - knew he ought not to leave it. He had already been attacked by a fire from the artillery in position beyond the Rummel buildings.

Major Weber, who was out on the crossroad leading northwest from the Low Dutch Road, had observed the movement of Stuart's column, headed by Chambliss and Jenkins, past the Stallsmith farm to the wooded crest behind Rummel's, and had reported it to Custer. Custer did, indeed, begin the movement. A portion of the Sixth Michigan and possibly of the Seventh also had begun to withdraw when Custer met Gregg coming on the field, and explained to him the situation - that the enemy was all around and preparing to push things. Gregg told him to remain where he was and that portion of the brigade which was moving away halted, countermarched and reoccupied its former position. The Fifth Michigan had not been withdrawn from the line in front, and Pennington's guns had never ceased to thunder their responses to the Confederate challenge.[12]

Custer says that the enemy opened upon him with a battery of six guns at 10 a.m. Stuart, on the contrary, claims to have left Gettysburg about noon. It is difficult to reconcile these two statements. A good deal of latitude may be given the word "about," but it is probable that the one puts the hour too early, while the other does not give it early enough; for of course, before Custer could be attacked, some portion of Stuart's command must have been upon the field. Official reports are often meagre, if not sometimes misleading, and must needs be reinforced by the memoranda and recollections of actual participants before the exact truth can be known.

Major Charles E. Storrs of the Sixth Michigan, who commanded a squadron, was sent out to the left and front of Custer's position soon after the brigade arrived upon the ground. He remained there several hours and was recalled about noon - he is positive it was later than twelve a.m. - to take position with the troops on the left of the battery. He states that the first shot was not fired until sometime after his recall, and he is sure it was not earlier than two o'clock.[13]

When Stuart left Gettysburg, as he says about noon, he took with him Chambliss's and Jenkins's brigades of cavalry and Griffin's battery. Hampton and Fitzhugh Lee were to follow; also Breathed's and McGregor's batteries, as soon as the latter had replenished their ammunition chests. Stuart moved two and a half miles out on the York turnpike, when he turned to the right by a country road that runs southeasterly past the Stallsmith farm. (This road intersects the Low Dutch Road, about three-fourths of a mile from where the latter crosses the Hanover pike.)

Turning off from this road to the right, Stuart posted the brigades of Jenkins and Chambliss and Griffin's battery on the commanding Cress's Ridge, beyond Rummel's and more than a mile from the position occupied by Custer. This movement was noticed by Major Weber, who with his detachment of the Sixth Michigan Cavalry was stationed in the woods northeast of Rummel's, where he could look out on the open country beyond, and he promptly reported the fact to Custer.

The first shot that was fired came from near the wood beyond Rummel's. According to Major McClellan, who was assistant adjutant general on Stuart's staff, this was from a section of Griffin's battery and was aimed by Stuart himself, he not knowing whether there was anything in his front or not. Several shots were fired in this way.

Major McClellan is doubtless right in this; that these shots were fired as feelers; but it is inconceivable that Stuart was totally unaware of the presence of any Federal force in his immediate front; that he did not know that there was stationed on the opposite ridge a brigade of cavalry and a battery. Gregg had been there the day before, and Stuart at least must have suspected, if he did not know, that he would find him there again. It is probable that he fired the shots in the hope of drawing out and developing the force he knew was there to ascertain how formidable it might be, and how great the obstacle in the way of his farther progress toward the rear of the Union lines. The information he sought was quickly furnished.

It was then that Custer put Pennington's battery in position, and the three sections of rifled cannon opened with a fire so fast and accurate that Griffin was speedily silenced and compelled to leave the field. Then there was a lull. I cannot say how long it lasted, but during its continuance, General Gregg arrived and took command in person. About this time also, it is safe to say that Hampton and Fitzhugh Lee came up and took position on the left of Chambliss and Jenkins. The Confederate line then extended clear across the Federal front, and was screened by the two patches of woods between Rummel's and the Stallsmith farm.

A battalion of the Sixth Michigan Cavalry, of which mine was the leading troop, was placed in support and on the left of Pennington's battery. This formed, at first, the short line of the "L" referred to in Custer's report, but it was subsequently removed farther to the right and faced in the same general direction as the rest of the line, where it remained until the battle ended. Its duty there was to repel any attempt that might be made to capture the battery.

The ground upon which these squadrons were stationed overlooked the plain, and the slightest demonstration in the open ground from either side was immediately discernible. From this vantage ground, it was possible to see every phase of the magnificent contest that followed. It was like a spectacle arranged for us to see. We were in the position of spectators at joust or tournament where the knights, advancing from their respective sides, charge full tilt upon each other in the middle of the field.

The lull referred to was like the calm that precedes the storm. The troopers were dismounted, standing "in place rest" in front of their horses, when suddenly there burst upon the air the sound of that terrific cannonading that preceded Pickett's charge. The earth quaked. The tremendous volume of sound volleyed and rolled across the intervening hills like reverberating thunder in a storm.

It was then between one and two o'clock. (Major Storrs says after two.) It was not long thereafter when General Custer directed Colonel Alger to advance and engage the enemy. The Fifth Michigan, its flanks protected by a portion of the Sixth Michigan on the left, by McIntosh's brigade on the right, moved briskly forward towards the wooded screen behind which the enemy was known to be concealed. In this movement, the right of regiment was swung well forward, the left somewhat "refused," so that Colonel Alger's line was very nearly at right angles with the left of Stuart's position.

As the Fifth Michigan advanced from field to field and fence to fence, a line of gray came out from behind the Rummel buildings and the woods beyond. A stubborn and spirited contest ensued. The opposing batteries filled the air with shot and shrieking shell. Amazing marksmanship was shown by Pennington's battery, and such accurate artillery firing was never seen on any other field. Alger's men with their eight-shotted carbines, forced their adversaries slowly but surely back, the gray line fighting well and superior in numbers, but unable to withstand the storm of bullets. It made a final stand behind the strong line of fences in front of Rummel's and a few hundred yards out from the foot of the slope whereon concealed by the woods, Stuart's reserves were posted.

While the fight was raging on the plain, Weber with his outpost was driven in. His two troops were added to the four already stationed on the left of Pennington's battery. Weber, who had been promoted to major but a few days before, was ordered by Colonel Gray to assume command of the battalion. As he took his place by my side in front of the leading troop, he said; "I have seen thousands of them over there," pointing to the front. "The country yonder is full of the enemy." He had observed all of Stuart's movements, and it was he who gave Custer the first important information as to what the enemy was doing; which information was transmitted to Gregg, and probably had a determining influence in keeping Custer on the field.

Weber was a born soldier, fitted by nature and acquirements for much higher rank than any he held. Although but 23 years of age, he had seen much service. A private in the Third Michigan Infantry in 1861, he was next battalion adjutant of the Second Michigan Cavalry, served on the staff of General Elliott in the southwest, and came home with Alger in 1862 to take a troop in the Sixth Michigan Cavalry. The valuable service rendered by him at Gettysburg was fitly recognized by Custer in his official report.

He was killed ten days later at Falling Waters while leading his squadron in a charge which was described by Kilpatrick as "the most gallant ever made." Anticipating a spirited fight, he was eager to have a part in it. "Bob," he said to me a few days before while marching through Maryland, "I want a chance to make one saber charge." He thought the time had come. His eye flashed and his face flushed as he watched the progress of the fight, fretting and chafing to be held in reserve when the bugle was summoning others to the charge.

The Fifth Michigan, holding the most advanced position, suffered greatly, Hampton having reinforced the Confederate line. Among those killed at this stage of the battle was Major Noah H. Ferry of the Fifth. Repeating rifles are not only effective but wasteful weapons as well, and Colonel Alger, finding that his ammunition had given out, felt compelled to retire his regiment and seek his horses. Seeing this, the enemy sprang forward with a yell. The Union line was seen to yield. The puffs of smoke from the muzzles of their guns had almost ceased. It was plain the Michigan men were out of ammunition and unable to maintain the contest longer.

On from field to field, the line of gray followed in exultant pursuit. Breathed and McGregor opened with redoubled violence. Shells dropped and exploded among the skirmishers, while thicker and faster they fell around the position of the reserves. Pennington replied with astonishing effect, for every shot hit the mark and the opposing artillerists were unable to silence a single Union gun. But still they came, until it seemed that nothing could stop their victorious career. "Men, be ready," said Weber. "We will have to charge that line." But the course of the pursuit took it toward the right, in the direction of Randol's battery where Chester was serving out canister with the same liberal hand displayed by Pennington's lieutenants; Clark, Woodruff and Hamilton.

Just then, a column of mounted men was seen advancing from the right and rear of the Union line. Squadron succeeded squadron until an entire regiment came into view, with sabers gleaming and colors gaily fluttering in the breeze. It was the Seventh Michigan, commanded by Colonel Mann. Gregg, seeing the necessity for prompt action, had given the order for it to charge. As the regiment moved forward and cleared the battery, Custer drew his saber, placed himself in front and shouted; "Come on you Wolverines!" The Seventh dashed into the open field and rode straight at the dismounted line which, staggered by the appearance of this new foe, broke to the rear and ran for its reserves. Custer led the charge half way across the plain, then turned to the left; but the gallant regiment swept on under its own leaders, riding down and capturing many prisoners.

There was no check to the charge. The squadrons kept on in good form. Every man yelled at the top of his voice until the regiment had gone perhaps five or six hundred yards straight towards the Confederate batteries, when the head of column was deflected to the left, making a quarter turn, and the regiment was hurled headlong against a post and rail fence that ran obliquely in front of the Rummel buildings. This proved for the time an impassable barrier. The squadrons coming up successively at a charge rushed pell-mell on each other and were thrown into a state of indescribable confusion, though the rear troops, without order or orders, formed left and right front into line along the fence, and pluckily began firing across it into the faces of the Confederates who, when they saw the impetuous onset of the Seventh thus abruptly checked, rallied and began to collect in swarms upon the opposite side.

Some of the officers leaped from their saddles and called upon the men to assist in making an opening. Among these were Colonel George G. Briggs, then adjutant, and Captain H.N. Moore. The task was a difficult and hazardous one, the posts and rails being so firmly united that it could be accomplished only by lifting the posts, which were deeply set, and removing several lengths at once. This was finally done however, though the regiment was exposed not only to a fire from the force in front, but to a flanking fire from a strong skirmish line along a fence to the right and running nearly at right angles with the one through which it was trying to pass.

While this was going on, Briggs's horse was shot and he found himself on foot, with three Confederate prisoners on his hands. With these, he started to the rear, having no remount. Before he could reach a place of safety, the rush of charging squadrons from either side had intercepted his retreat. In the melee that followed, two of his men ran away, the other undertook the duty of escorting his captor back to the Confederate lines. The experiment cost him his life, but the plucky adjutant, although he did not run away, lived to fight again on many another day.

In the meantime, through the passage way thus effected, the Seventh moved forward, the center squadron leading, and resumed the charge. The Confederates once more fell back before it. The charge was continued across a plowed field to the front and right, up to and past Rummel's to a point within 200 or 300 yards of the Confederate battery. There, another fence was encountered, the last one in the way of reaching the battery, the guns of which were pouring canister into the charging column as fast as they could fire. Two men, privates Powers and Inglede of Captain Moore's troop, leaped this fence and passed several rods beyond. Powers came back without a scratch, but Inglede was severely wounded. These two men were certainly within 200 yards of the Confederate cannon.

But, seeing that the enemy to the right had thrown down the fences and was forming a column for a charge, the scattered portions of the Seventh began to fall back through the opening in the fence. Captain Moore, in whose squadron sixteen horses had been killed, retired slowly, endeavoring to cover the retreat of the dismounted men, but taking the wrong direction, came to the fence about 100 yards above the opening just as the enemy's charging column struck him.

Glancing over his shoulder, he caught the gleam of a saber thrust from the arm of a sturdy Confederate. He ducked to avoid the blow, but received the point in the back of his head. At the same time, a pistol ball crashed through his charger's brain and the horse went down, Moore's leg under him. An instant later, Moore avenged his steed with the last shot in his revolver, and the Confederate fell dead at his side. Some dismounted men of the Thirteenth Virginia Cavalry took Moore prisoner and escorted him back to the rear of their battery, from which position, during the excitement that followed, he made his escape.

But now Alger, who when his ammunition gave out hastened to his horses, had succeeded in mounting one battalion, commanded by Major L.S. Trowbridge, and when the Ninth and Thirteenth Virginia struck the flank of the Seventh Michigan, he ordered that officer to charge and meet this new danger. Trowbridge and his men dashed forward with a cheer, and the enemy in their turn were put to flight. Past the Rummel buildings, through the fields, almost to the fence where the most advanced of the Seventh Michigan had halted, Trowbridge kept on. But he too was obliged to retire before the destructive fire of the Confederate cannon, which did not cease to belch forth destruction upon every detachment of the Union cavalry that approached near enough to threaten them. The major's horse was killed, but his orderly was close at hand with another and he escaped. When his battalion was retiring, it also was assailed in flank by a mounted charge of the First Virginia Cavalry, which was met and driven back by the other battalion of the Fifth Michigan led by Colonel Alger.

Then, as it seemed, the two belligerent forces paused to get their second breath. Up to that time, the battle had raged with varying fortune. Victory that appeared about to perch first on one banner, and then on the other, held aloof as if disdaining to favor either. The odds indeed had been rather with the Confederates than against them, for Stuart managed to outnumber his adversary at every critical point, though Gregg forced the fighting, putting Stuart on his defense, and checkmating his plan to fight an offensive battle. But the wily Confederate had kept his two choicest brigades in reserve for the supreme moment, intending then to throw them into the contest and sweep the field with one grand, resistless charge.

All felt that the time for this effort had come, when a body of mounted men began to emerge from the woods on the left of the Confederate line, northeast of the Rummel buildings, and form column to the right as they debouched into the open field. Squadron after squadron, regiment after regiment, orderly as if on parade, came into view, and successively took their places.

Then Pennington opened with all his guns. Six rifled pieces, as fast as they could fire, rained shot and shell into that fated column. The effect was deadly. Great gaps were torn in that mass of mounted men, but the rents were quickly closed. Then, they were ready. Confederate chroniclers tell us there were two brigades - eight regiments – under their own favorite leaders. In the van floated a stand of colors. It was the battle flag of Wade Hampton, who with Fitzhugh Lee was leading the assaulting column. In superb form with sabers glistening, they advanced. The men on foot gave way to let them pass. It was an inspiring and an imposing spectacle that brought a thrill to the hearts of the spectators on the opposite slope. Pennington double-shotted his guns with canister, and the head of the column staggered under each murderous discharge. But still it advanced, led on by an imperturbable spirit that no storm of war could cow.

Major Luther Trowbridge

Meantime, the Fifth Michigan had drawn aside a little to the left, making ready to spring. McIntosh's squadrons were in the edge of the opposite woods. The Seventh was sullenly retiring with faces to the foe. Weber and his battalion and the other troops of the Sixth were on edge for the fray, should the assault take the direction of Pennington's battery which they were supporting.

On and on, nearer and nearer, came the assaulting column, charging straight for Randol's battery. The storm of canister caused them to waver a little, but that was all. A few moments would bring them among Chester's guns who like Pennington's lieutenants, was still firing with frightful regularity, as fast as he could load. Then Gregg rode over to the First Michigan and directed Town to charge. Custer dashed up with similar instructions, and as Town ordered sabers to be drawn, placed himself by his side in front of the leading squadron.

With ranks well closed, with guidons flying and bugles sounding, the grand old regiment of veterans, led by Town and Custer, moved forward to meet that host, outnumbering it three to one. First at a trot, then the command to charge rang out, and with gleaming saber and flashing pistol, Town and his heroes were hurled right in the teeth of Hampton and Fitzhugh Lee. Alger, who with the Fifth had been waiting for the right moment, charged in on the right flank of the column as it passed, as did some of McIntosh's squadrons on the left. One troop of the Seventh, led by Lieutenant Dan Littlefield, also joined in the charge.

Then it was steel to steel. For minutes - and for minutes that seemed like years - the gray column stood and staggered before the blow; then yielded and fled. Alger and McIntosh had pierced its flanks, but Town's impetuous charge in front went through it like a wedge, splitting it in twain and scattering the Confederate horsemen in disorderly rout back to the woods from whence they came.

During the last melee, the brazen lips of the cannon were dumb. It was a hand-to-hand encounter between the Michigan men and the flower of the southern cavaliers, led by their favorite commanders. Stuart retreated to his stronghold, leaving the Union forces in possession of the field.

The rally sounded, the lines were reformed, the wounded were cared for and everything was made ready for a renewal of the conflict. But the charge of the First Michigan ended the cavalry fighting on the right at Gettysburg. Military critics have pronounced it the finest cavalry charge made during that war.

Custer's brigade lost one officer (Major Ferry) and 28 men killed; 11 officers and 112 men wounded; 67 men missing; total loss, 219. Gregg's division lost one man killed; 7 officers and 19 men wounded; 8 men missing; total, 35. In other words, while Gregg's division, two brigades, lost 35, Custer's single brigade suffered a loss of 219. These figures apply to the fight on July 3rd only. The official figures show that the brigade, during the three days, July 1, 2 and 3, lost 1 officer and 31 men killed; 13 officers and 134 men wounded; 78 men missing; total, 257.[14]

Gettysburg. Today at Gettysburg, monuments to both Union and Confederate troops are visible on the battlefield. The Gettysburg National Military Park preserves one of the world's largest collections of outdoor sculptures with close to 1,400 statues, sculptures, markers and tablets standing where men fought — memorials to the sacrifices at Gettysburg.

For more than twenty years after the close of the Civil War, the part played by Gregg, Custer and McIntosh and their brave followers in the Battle of Gettysburg received but scant recognition. Even the maps prepared by the Corps of Engineers stopped short of Cress's Ridge and Rummel's fields. History was practically silent upon the subject, and had not the survivors of those commands taken up the matter, there might have been no record of the invaluable services which the Second Cavalry

Division and Custer's Michigan Brigade rendered at the very moment when a slight thing would have turned the tide of victory the other way. In other words, the decisive charge of Colonel Town and his Michiganders coincided in point of time with the failure of Pickett's assault upon the center, and was a contributing cause in bringing about the latter result.

About the year 1884, a monument was dedicated on the Rummel farm which was intended to mark as nearly as possible the exact spot where Gregg and Custer crossed swords with Hampton and Fitzhugh Lee in the final clash of the cavalry fight. This monument was paid for by voluntary contributions of the survivors of the men who fought with Gregg and Custer. Colonel George Gray of the Sixth Michigan alone contributed four hundred dollars. Many others were equally liberal. On that day, Colonel Brooke-Rawle of Philadelphia, who served in the Third Pennsylvania Cavalry of Gregg's division, delivered an address upon the "Cavalry Fight on the Right Flank at Gettysburg."

It was an eloquent tribute to Gregg and his Second Division and to the Michigan Brigade, though like a loyal knight, he claimed the lion's share of the glory for his own and placed chaplets of laurel upon the brow of his ideal hero of Pennsylvania rather than upon that of "Lancelot, or another." In other words, he did not estimate Custer's part at its full value, an omission for which he subsequently made graceful and honorable acknowledgment. In this affair, there were honors enough to go around.

Subsequently, General Luther S. Trowbridge of Detroit, who was an officer in the Fifth Michigan Cavalry, who like Colonel Brooke-Rawle fought most creditably in the cavalry fight on the right, wrote a paper on the same subject which was read before the Michigan commandery of the Loyal Legion. This very fitly supplemented Colonel Brooke-Rawle's polished oration. In the year 1889, another monument erected by the State of Michigan on the Rummel farm, and but a hundred yards or such a matter from the other, was dedicated.

The writer of these "Recollections" was the orator of the occasion, and the points of his address are contained in the narrative which constitutes this chapter. Those three papers and others written since that time, notably one by General George B. Davis, judge advocate general, U.S.A., and one by Captain Miller of the Third Pennsylvania Cavalry, have brought the cavalry fight at Gettysburg into the limelight so that there is no longer any pretext for the historian or student of the history of the Civil War to profess ignorance of the events of that day, which reflect so much luster on the cavalry arm of the service.

To illustrate the point made in these concluding paragraphs that the part taken by the cavalry on the right is at last understood and acknowledged, the following extract from an address given before the students of the Orchard Lake Military Academy by General Charles King, the gifted author of "The Colonel's Daughter," and many other writings, is herein quoted. General King is himself a cavalry officer with a brilliant record in the army of the United States. In that address to the students on "The Battle of Gettysburg," he said:

"And so, just as Gettysburg was the turning point of the Great War, so to my thinking was the grapple with and overthrow of Stuart on the fields of the Rummel farm the turning point of Gettysburg. Had he triumphed there; had he cut his way through or over that glorious brigade of Wolverines and come sweeping all before him down among the reserve batteries and ammunition trains, charging furiously at the rear of our worn and exhausted infantry even as Pickett's devoted Virginians assailed their front, no man can say what scenes of rout and disaster might not have occurred."

"Pickett's charge was the grand and dramatic climax of the fight because it was seen of all men. Stuart's dash upon the Second Division far out on the right flank was hardly heard of for years after. It would have rung the world over but for the Michigan men. Pennsylvania and New Jersey, New York and the little contingent of Marylanders had been fighting for days, were scattered, dismounted and exhausted when the plumes of Stuart came floating out from the woods of the Stallsmith farm, Hampton and Fitzhugh Lee at his back. It was Custer and the Wolverines who flew like bulldogs straight at the throat of the foes; who blocked his headlong charge; who pinned him to the ground while like wolves their comrade troops rushed upon his flanks."

"It may be, perhaps an out-cropping of the old trooper spirit now, but as I look back upon the momentous four years' struggle, with all its lessons of skill and fortitude and valor incomparable, it seems to me that, could I have served in only one of its great combats, drawn saber in just one supreme crisis on whose doubtful issue hung trembling the fate of the whole Union, I would beg to live that day over again and to ride with Gregg and McIntosh and Custer; to share in the wild, fierce charge of the Michigan men; to have my name go down to posterity with those of Alger and Kidd, Town and Trowbridge, Briggs and gallant Ferry, whose dead hand gripped the saber hilt and the very grave."

"To have it said that I fought with the old Second Division of the cavalry corps that day when it went and grappled and overwhelmed the foe in the full tide of his career, at the very climax of the struggle, and hurled him back to the banks of the Rubicon of the Rebellion, to cross it then and there for the last time, to look his last upon the green hills of Maryland - nevermore to vex our soil until, casting away the sword, he could come with outstretched hand and be hailed as friend and brother."

FROM GETTYSBURG TO FALLING WATERS

WHEN the Battle of Gettysburg was ended and the shadows of night began to gather upon the Rummel fields, the troopers of the Michigan Cavalry Brigade had a right to feel that they had acted well their parts, and contributed their full share to the glory and success of the Union arms. They had richly earned a rest, but were destined not to obtain it until after many days of such toil and hardship as to surpass even the previous experiences of the campaign.

After a brief bivouac on the battlefield, the brigade was moved to the Baltimore pike, whence at daybreak it marched to the vicinity of Emmittsburg. There, on the morning of July 4, the two brigades of the Third Division reunited. The First Brigade, under the lamented Farnsworth, it will be remembered had been engaged the previous day upon the left flank near "Round Top," under the eye of the division commander.

Farnsworth, the gallant young officer who had been a brigadier general but four days, had been killed while leading a charge against infantry behind stone walls. His brigade was compelled to face infantry because all of the Confederate cavalry had been massed under Stuart against Meade's right. It was intended that Custer should report to Kilpatrick on the left flank, but as we have seen, he was providentially where he was most needed, and where his presence was effective in preventing disaster.

The charge in which Farnsworth lost his life was ordered by Kilpatrick and was unquestionably against the former's judgment. But he was too brave a man and too conscientious to do anything else than obey orders to the letter. His courage had been put to the proof in more than a score of battles. As an officer in the Eighth Illinois Cavalry and as an aid on the staff of General Pleasonton, Chief of Cavalry, he had won such deserved distinction that he, like Custer, was promoted from captain to brigadier general on June 28th and assigned to command of the First Brigade of Kilpatrick's division when Custer took the Second.

This was done in spite of the fact that he was not a graduate of the military academy or even an officer of the regular army. I knew him before the war when he was a student in the University of Michigan, and a more intrepid spirit than he possessed never resided within the breast of man. It was but a day, it might be said, that he had worn his new honors. He was proud, ambitious, spirited, loyal, brave, true as steel to his country and his convictions of duty, and to his own manhood.

He did not hesitate for one moment. Drawing his saber and placing himself at the head of his command, he led his men to the inevitable slaughter and boldly went to his own death. It was a pity to sacrifice such an officer and such men as followed him inside the Confederate lines. The charge was one of the most gallant ever made, though barren of results. The little force came back shattered to pieces and without their leader. The cavalry corps had lost an officer whose place was hard to fill. Had he lived, the brave young Illinoisan might have been another Custer. He had all the qualities needed to make a great career - youth, health, a noble physique, courage, patriotism, ambition, ability and rank. He was poised, like Custer, and had discretion as well as dash.

They were a noble pair, and nobly did they justify the confidence reposed in them. One lived to court death on scores of battlefields, winning imperishable laurels in them all; the other was cut down in the very beginning of his brilliant career, but his name will forever be associated with what is destined to be in history the most memorable battle of the war, and the one from which is dated the beginning of the downfall of the Confederate cause and the complete restoration of the Union. Farnsworth will not be forgotten as long as a grateful people remember the name and the glory of Gettysburg.

John Elon Farnsworth. After the defeat of J.E.B. Stuart's Confederate cavalry on July 3, the third day of the Battle of Gettysburg, General Kilpatrick, commanding the 3rd Division, ordered Farnsworth to make a charge with his brigade against Confederate positions south of the Devil's Den. Farnsworth initially balked, arguing there was no hope of success, and only agreed when Kilpatrick accused him of cowardice. The charge was repulsed with heavy losses, and Farnsworth was killed. Kilpatrick received heavy criticism for ordering the charge, but no official action was taken against him.

Although General Judson Kilpatrick had been in command of the division since the 30th of June, at Hanover, many of the Michigan men had never set eyes upon him until that morning, and there was much curiosity to get a sight of the already famous cavalryman. He had begun to be a terror to foes, and there was a well grounded fear that he might become a menace to friends as well. He was brave to rashness, capricious, ambitious, reckless in rushing into scrapes, and generally full of expedients in getting out, though at times he seemed to lose his head entirely when beset by perils which he himself had invited. He was prodigal of human life, though to do him justice he rarely spared himself. While he was not especially refined in manners and in conversation, he had an intellect that would at times emit flashes so brilliant as to blind those who knew him best to his faults. He was the very type of one of the wayward cavaliers who survived the death of Charles the First, to shine in the court of Charles the Second. He was a ready and fluent speaker - an orator, in fact - and had the gift of charming an audience with his insinuating tongue.

As closely as I can from memory, I will draw a pen sketch of him as he appeared at that time: Not an imposing figure as he sat with a jaunty air upon his superb chestnut horse, for he was of slight build though supple and agile as an athlete; a small, though well knit form, dressed in a close-fitting and natty suit of blue; a blouse with the buttons and shoulder straps of a brigadier-general; the conventional boots and spurs and saber; a black hat with the brim turned down on one side, up on the other, in a way affected by himself, which gave to the style his own name. This completed his uniform - not a striking or picturesque one in any respect. Save for the peculiar style of hat, there was nothing about it to distinguish him from others of like rank. But his face was a marked one, showing his individuality in every line. A prominent nose, a wide mouth, a firm jaw, thin cheeks set off by side whiskers rather light in color, and eyes that were cold and lusterless, but searching - these were the salient characteristics of a countenance that once seen was never forgotten. His voice had a peculiar, piercing quality, though it was not unmusical in sound.

In giving commands, he spoke in brusque tones and in an imperious manner. It was not long until every man in the division had seen him and knew him well. In a few days, he had fairly earned the soubriquet "Kill Cavalry," which clung to him until he left for the west. This was not because men were killed while under his command, for that was their business and every trooper knew that death was liable to come soon or late while he was in the line of duty, but for the reason that so many lives were sacrificed by him for no good purpose whatever.

General Judson Kilpatrick Was Nicknamed "Kill Cavalry" for using tactics in battle with a reckless disregard for lives of soldiers under his command. Summing up Kilpatrick in 1864, General Sherman said "I know that Kilpatrick is a hell of a damned fool, but I want just that sort of man to command my cavalry on this expedition."

Well, on the morning of the Fourth, General Kilpatrick sent an order to regimental commanders to draw three days' rations and be prepared for a protracted absence from the army, as we were to go to the right and rear of Lee to try and intercept his trains, and in every way to harass his retreating columns as much as possible. We were all proud of our new commanders, for it was evident that they were fighting men and that while they would lead us into danger, if we survived it there would be left the consciousness of having done our duty, and the credit of accomplishing something for the cause.

It must also be said that a strong feeling of "pride in the

corps" had taken root. Men were proud that they belonged to Kilpatrick's division and to Custer's brigade, for it must not be supposed that the above estimate of the former is based upon what we knew of him at that time. We were under him for a long time after that. This was the first day that we felt the influence of his immediate presence. When it was known that Kilpatrick was to lead a movement to the enemy's rear, all felt that the chances were excellent for the country to hear a good deal about our exploits within the next few days, and nobody regretted it.

But before the start, it began to rain in torrents. It has been said that a great battle always produces rain. My recollection is not clear as to the other battles, but I know that the day after Gettysburg, the flood gates of heaven were opened, and as the column of cavalry took its way towards Emmittsburg, it was deluged. It seemed as if the firmament were an immense tank, the contents of which were spilled all at once. Such a drenching as we had! Even heavy gum coats and horsehide boots were hardly proof against it. It poured and poured, the water running in streams off the horses' backs, making of every rivulet a river and of every river and mountain stream a raging flood.

But Lee was in retreat, and rain or shine, it was our duty to reach his rear, so all day long we plodded and plashed along the muddy roads towards the passes in the Catoctin and South Mountains. It was a tedious ride for men already worn out with incessant marching and the fatigues of many days. It hardly occurred to the tired trooper that it was the anniversary of the nation's natal day. There were no fireworks, and enthusiasm was quenched not by the weather only but by the knowledge that the Confederate army, though repulsed, was not captured. The news of Grant's glorious victory in the west filled every heart with joy of course, but the prospect of going back into Virginia to fight the war over again was not alluring. But possibly that might not be our fate. Vigorous pursuit might intercept Lee on this side of the Potomac. Every trooper felt that he could endure wet and brave the storm to aid in such strategy, and all set their faces to the weather and rode, if not cheerfully, at least patiently forward in the rain.

I have said that on that memorable Fourth of July, there were no fireworks. That was a mistake. The pyrotechnic display was postponed until a late hour, but it was an interesting and exciting exhibition, as all who witnessed it will testify. It was in the night and darkness lent intensity to the scene.

Toward evening, the flood subsided somewhat, though the sky was overcast with wet looking clouds, and the swollen and muddy streams that ran along and across our pathway fretted and frothed like impatient coursers under curb and rein. Their banks could hardly hold them. During the afternoon and evening, the column was climbing the South Mountain. A big Confederate wagon train was going through the gap ahead of us. If we could capture that, it would be making reprisal for some of Stuart's recent work in Maryland.

Toward midnight, we were nearing the top, marching along the narrow defile, the mountain towering to the right and sloping off abruptly to the left, when the boom of a cannon announced that the advance guard had encountered the enemy. The piece of artillery was planted in the road, at the summit near the Monterey House, and was supported by the Confederate rear guard, which at once opened fire with their carbines. It was too dark to distinguish objects at any distance, the enemy was across the front and no one could tell how large a force it might be.

The First Michigan had been sent to the right early in the evening to attack a body of the enemy, hovering on the right flank in the direction of Fairfield, and had a hard fight in which Captain Elliott and Lieutenant McElhenny, two brave officers, were killed. The Fifth and Sixth were leading and at once dismounted and deployed as skirmishers. Generals Kilpatrick and Custer rode to the place where the line was forming and superintended the movement. The Sixth, under Colonel Gray, was on the right of the line, the road to its left. At least the portion of the regiment to which my troop belonged was in that position. I think, perhaps, a part of the regiment was across the road. The Fifth formed on the left; the First and Seventh in reserve, mounted. There is a good deal of guess work about it, for in the darkness one could not tell what happened except in his immediate neighborhood.

Antietam Cemetery in foreground, South Mountain in background. Gap shows location of Signal Station during the battle.

The order "Forward," finally came, and the line of skirmishers advanced up the slope, a column of mounted men following in the road, ready to charge when opportunity offered. Soon we encountered the Confederate skirmishers, but could locate them only by the flashes of their guns. The darkness was intense, and in a few moments we had plunged into a dense thicket, full of undergrowth, interlaced with vines and briars, so thick that it was difficult to make headway at all. More than once a trailing vine tripped me up, and I fell headlong. To keep up an alignment was out of the question. One had to be guided by sound and not by sight.

The force in front did not appear to be formidable in

numbers, but had the advantage of position, and was on the defensive in a narrow mountain pass where numbers were of little avail. We had a large force, but it was strung out in a long column for miles back, and it was possible to bring only a few men into actual contact with the enemy, whatever he might be. This last was a matter of conjecture, and Kilpatrick doubtless felt the necessity of moving cautiously, feeling his way until he developed what was in his front. To the right of the road, had it not been for the noise and the flashing of the enemy's fire, we should have wandered away in the darkness and been lost.

The Confederate skirmishers were driven back across a swollen stream spanned by a bridge. The crossing at this point was contested fiercely, but portions of the Fifth and Sixth finally forced it and then the whole command crossed over. In the meantime, the rumbling of wagon wheels could be heard in the road leading down the mountain. It was evident we were being detained by a small force striving to hold us there while the train made its escape.

A regiment was ordered up mounted to make a charge. I heard the colonel giving his orders. "Men," he said, "use the saber only; I will cut down any man who fires a shot." This was to prevent shooting our own men in the melee, and in the darkness. Inquiring, I learned it was the First (West) Virginia Cavalry. This regiment, which belonged in the First Brigade, had been ordered to report to Custer. At the word, the gallant regiment rushed like the wind down the mountain road, yelling like troopers as they were, and good ones too, capturing everything in their way.

This charge ended the fighting for that night. It was one of the most exciting engagements we ever had, for while the actual number engaged was small and the casualties were not great, the time, the place, the circumstances, the darkness, the uncertainty, all combined to make "the midnight fight at Monterey" one of unique interest. General Custer had his horse shot under him, which it was said and I have reason to believe was the seventh horse killed under him in that campaign. The force that resisted us did its duty gallantly, though it had everything in its favor. They knew what they had in their front, we did not. Still, they failed of their object, which was to save the train. That, we captured after all. The Michigan men brushed the rear guard out of the way, the First Virginia gave the affair the finishing touch.

The fight over, men succumbed to fatigue and drowsiness. I had barely touched the saddle before I was fast asleep, and did not awake until daylight and then looking around, could not see a man that I recognized as belonging to my own troop. As far as the eye could reach, both front and rear, was a moving mass of horses with motionless riders all wrapped in slumber. The horses were moving along with drooping heads and eyes half closed. Some walked faster than others, and as a consequence would gradually pull away from their companions through the column in front; others would fall back.

So it came to pass that few men found themselves in the same society in the morning with which they started at midnight. As for myself, I awoke to wonder where I was and what had become of my men. Not one of them could I see. My horse was a fast walker, and I soon satisfied myself that I was in advance of my troop, and when the place designated for the division to bivouac was reached, dismounted and awaited their arrival. Some of them did not come up for an hour, and they were scattered about among other commands, in squads, a few in a place. It was seven o'clock before we were all together once more.

Then we had breakfast, and the men had a chance to look the captures over and quiz the prisoners. The wagons were soon despoiled of their contents and such stuff as was not valuable or could not be transported was burned. Among the prisoners was Colonel Davis of the Tenth Virginia Cavalry, who claimed that he led the charge against our position on the third. He expressed himself very freely as having had enough, and said, "This useless war ought to be ended at once."

During the day, Stuart's cavalry appeared on our flank and we pushed on to Cavetown, thence to Boonsborough, harassed all the way by the enemy. We were now directly on Lee's path to the Potomac. At Smithburg, there was quite a skirmish in which the Sixth had the duty of supporting the battery. My troop, deployed as skirmishers along the top of a rocky ridge, was forgotten when the division moved away after dark, and we lay there for an hour within sight of the Confederate camp until suspecting something wrong, I made a reconnoissance and discovered that our command had gone. I therefore mounted the men and followed the trail which led toward Boonsborough. At the latter place, Kilpatrick turned over his prisoners and captured property.

On the 6th, along in the afternoon, we arrived in the vicinity of Hagerstown. The road we were on enters the town at right angles with the pike from Hagerstown to Williamsport, on reaching which we turned to the left, the position being something like the following diagram:

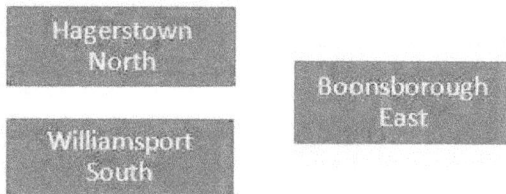

```
Hagerstown
North
                        Boonsborough
                        East
Williamsport
South
```

Lee's reserve wagon trains were at Williamsport under General Imboden. From Hagerstown to Williamsport was about five miles. We had Stuart's cavalry in our front, Lee's whole army on our right, and only five miles to our left the tempting prize which Kilpatrick was eager to seize. Besides, it was necessary for Lee to reach Williamsport in order to secure a crossing of the Potomac River.

The advance of his army reached Hagerstown simultaneously with ourselves, but the skirmishers of the First Brigade drove them back to the northward, and then the Michigan Brigade passed through and turned southward on the pike toward Williamsport. The Fifth Michigan had the advance and the Sixth the rear. The latter regiment had hardly more than turned in the new direction when the boom of a cannon in front told the story that the battle had begun.

Confederate dead by a road near Hagerstown. Hagerstown's strategic location at the border between the North and the South made it a primary staging area and supply center for four major campaigns during the Civil War. Throughout the war, physicians and citizens of Hagerstown gave assistance or aid to men from both North and South at a number of locations, including the Franklin Hotel, Washington House, Hagerstown Male Academy, Key-Mar College, and private residences.

General Kilpatrick had been attending to matters in Hagerstown. It was evident that there was considerable force there and that it was constantly augmenting. The opening gun at Williamsport called his attention to a new danger. It looked as though he had deliberately walked into a trap. In a moment I saw him coming, dashing along the flank of the column. He was urging his horse to its utmost speed. In his hand, he held a small riding whip with which he was touching the flank of his charger as he rode. His face was pale. His eyes were gazing fixedly to the front and he looked neither to the right nor to the left. The look of anxiety on his countenance was apparent. The sound of cannon grew louder and more frequent; we were rushed rapidly to the front. The First Brigade followed and to the officer in command of it was assigned the task of holding back Lee's army while the Michigan Brigade tried titles with Imboden. Buford, with the First Cavalry Division, was fighting Stuart's cavalry to the left towards Boonsborough, and on him it depended to keep open the only avenue of escape from the position in which Kilpatrick found himself.

In a little time, the two brigades were fighting back to back, one facing north and the other south, and each having more than it could attend to. Pretty soon, we arrived on the bluff overlooking Williamsport. Imboden's artillery had the exact range and were pouring shell into the position where the brigade was trying to form.

Just before arriving at the point where we were ordered to turn to the right through an opening in a rail fence into a field, Aaron C. Jewett, acting adjutant of the regiment, rode along the column delivering the order from the colonel. During the Gettysburg campaign, Jewett had been acting adjutant and would have received his commission in a short time. His modest demeanor and affable manners had won the hearts of all his comrades. He had made himself exceedingly popular, as well as useful, and was greatly beloved in the regiment. When he delivered the order, the pallor of his countenance was noticeable. There was no tremor, no shrinking, no indication of fear; he was intent upon performing his duty; gave the order, and turning, galloped back to where the shells were flying thick and fast.

When I arrived at the gap in the fence, he was there; he led the way into the field; told me where to go in; there was no trepidation on his part but still that deathly pallor. As we passed into the field, a shell exploded directly in front of us. It took a leg off a man in troop H, which preceded us and had dismounted to fight on foot, and I saw him hopping around on his one remaining limb and heard him shriek with pain. A fragment of the same shell took a piece off the rim of Lieutenant E.L. Craw's hat. He was riding at my side. I believe it was the same shell that killed Jewett. He had left me to direct the next troop in order, and a fragment of one of these shells struck him in the throat and killed him instantly. As I moved rapidly forward after getting into the field, I did not see him again, and did not know he was killed until after dark, when we had succeeded in making our escape by a very narrow chance.

We were moved well over to the right - all the time under a furious fire of artillery - and kept there until almost dark, fighting all the time with the troops that were pushed out from Williamsport. In the meantime, the firing and yelling in rear could be heard distinctly and it seemed that at any moment, the little force was to be closed in on and captured. Finally, just after dark, it was withdrawn. Those on the right of the road - the First and Sixth - the Fifth and Seventh being to the left, were obliged to reach and cross the pike to make their escape. Weber stealthily withdrew the battalion. He was the last man to leave the field. When we were forming in the road after rallying the skirmishers, the enemy was in plain sight only a little way toward Hagerstown, and it seemed as if one could throw a stone and hit them. We expected they would charge us, but they did not, and probably the growing darkness prevented it. In fact, there was manifest a disposition on their part to let us alone if we would not molest them.

We then marched off into a piece of woods, and the regiment having all reunited learned - those who had not known of it before - of Jewett's death. His body was still where it fell. The suggestion was made to go and recover it. Weber and his men made an attempt to do so, but by that time the enemy had come up and taken possession of the field. This was a terrible blow to all, to be obliged to leave the body of a beloved comrade; to be denied the privilege of aiding in placing him in a soldier's grave and performing the last offices of affection for a fallen friend.

The death of Jewett was a blow to the regiment, the more severe because he was the first officer killed up to that time. A portion of the regiment had been roughly handled on the evening of July 2nd at Hunterstown - where Thompson and Ballard were wounded - and the latter taken prisoner. A number of the rank and file were in the list of killed, wounded and missing. Enough had been seen of war to bring to all a realization of its horrors. Death was a familiar figure, yet Jewett's position as adjutant had brought him into close relations with both officers and men, and his sudden death was felt as a personal bereavement. It was like coming into the home and taking one of the best beloved of the household.

After getting out of the Williamsport affair, most of the night was taken up in marching, and on the morning of the 7th, the brigade was back in Boonsborough, where remaining in camp all day, it obtained a much needed rest, though the Fourth of July rain storm was repeated. Lee's army had reached the Potomac, and not being able to cross by reason of the high water was entrenching on the north side. Meade's army was concentrating in the vicinity but seemed in no hurry about it. During the day, some heavy siege guns, coming down the mountain road, passed through Boonsborough going to the front.

A big battle was expected to begin at any moment, and we wondered why there was so much deliberation when Lee's army was apparently in a trap with a swollen river behind it. It did not seem possible that he would be permitted to escape into Virginia without fighting a battle. To the cavalry of Kilpatrick's division, which had been marching and countermarching over all the country between the South Mountain and the Potomac River, the delay was inexplicable. Every trooper believed that the Army of the Potomac had the Confederacy by the throat at last, and that vigorous and persistent effort would speedily crush the life out of it.

But no battle took place, and on the morning of the 8th, Stuart's cavalry, which was now covering Lee's front, was attacked in front of Boonsborough by Buford and Kilpatrick, and a hard battle resulted. Most of the fighting was done dismounted, the commands being deployed as skirmishers. Custer's brigade occupied the extreme left of the line, and I think the Sixth the left of the brigade. The enemy was also on foot, though many mounted officers could be seen on their line. We had here a good opportunity to test the qualities of the Spencer carbines and armed as we were, we proved more than a match for any force that was encountered.

Antietam Bridge on the Boonsborough pike with stone house on the east side, by Alexander Gardner, 1821-1882.

The firing was very sharp at times, and took on the character of skirmishing, the men taking advantage of every cover that presented itself. The Confederates were behind a stone fence, we in a piece of woods along a rail fence which ran along the edge of the timber. Between was an open field. Several times they attempted to come over the stone wall and advance on our position, but each time were driven back. Once, an officer jumped up on the fence and tried to wave his men forward. A shot from a Spencer brought him headlong to the ground, and after that, no one had the temerity to expose himself in that way.

At this stage of the battle (it must have been about eleven o'clock in the forenoon), a singular thing happened. It is one of those numberless incidents that do not appear in official reports, and which give to individual reminiscences their unique interest. An officer, dressed in blue, with the regulation cavalry hat, riding a bay horse which had the look of a thoroughbred, rode along in rear of our line with an air of authority, and with perfect coolness said, as he passed from right to left, "General Kilpatrick orders that the line fall back rapidly." The order was obeyed promptly, though it struck us as strange that such a strong position should be given up without a struggle. We had not been under Kilpatrick long enough to recognize all the members of his staff on sight, and it did not occur to anyone at the time to question the fellow's authority or make him show his credentials.

The line left the woods and retreated to a good defensive position on a ridge of high ground facing the woods, the enemy meantime advancing with a yell to the timber we had abandoned. Then it was learned that Kilpatrick had given no such order, but the "staff officer" had disappeared, and when we came to think about it, nobody could describe him very closely. He had seemed to flit along the line, giving the order but stopping nowhere, and leaving no very clear idea as to how he looked. There is but little doubt that he was an audacious Confederate, probably one of Stuart's scouts clothed in Federal uniform, who made a thorough tour of inspection of our line, and then, after seeing us fall back, very likely led his own line to the position which he secured by this daring stratagem. The Confederates were up to such tricks, and occasionally the Yankees were smart enough to give them a Roland for their Oliver.

It was presently necessary to advance and drive the enemy out of the woods, which was done in gallant style, the whole line joining. This time there was no stopping, but the pursuit was kept up for several miles. I can hear gallant Weber's voice now as he shouted, "Forward, my men," and leaping to the front led them in the charge.

The Fifth Michigan was to our right, and Colonel Alger, who was in command, was wounded in the leg and had to leave the field. We did not see him again for some time, the command devolving upon Lieutenant Colonel Gould who in turn was himself wounded a day or two later, and Major Luther S. Trowbridge, who did such gallant fighting at Gettysburg, succeeded to the command.

From the night of the 8th to the morning of the 11th, there was an interval of quietude. The cavalry was waiting and watching for Lee or Meade to do something, and to the credit of the Union troopers, it must be said that they were eager for the conflict to begin, believing as they did that the war ought to end in a day.

July 11th, early in the morning, an attack was made on the lines around Hagerstown, which developed a hornets' nest of sharpshooters armed with telescopic rifles, who could pick a man's ear off half-a-mile away. The bullets from their guns had a peculiar sound, something like the buzz of a bumblebee, and the troopers' horses would stop, prick up their ears and gaze in the direction whence the hum of those invisible messengers could be heard. Unable to reach them mounted, we finally deployed dismounted along a staked rail fence. The Confederates were behind trees and shocks of grain at least half-a-mile away. They would get the range so accurately that it was dangerous to stand still a moment. It was possible however, to dodge the bullets by observing the puffs of smoke from their guns.

The distance was so great that the puff was seen some seconds before the report was heard, and before the arrival of the leaden missile. By moving to the right or left, the shot could be avoided, which in many cases was so accurately aimed as to have been fatal had it been awaited. Once, I was slow about moving. The scamp in my immediate front had evidently singled me out and was sending them in so close as to make it sure that he was taking deadly aim. I took my eye off his natural fortress for an instant when he fired, and before I could jump, the ball struck a rail in front of me, and passing through the rail, fell to the ground at my feet.

Most of the men were content to keep behind the fence and try and give the Confederates as good as they sent, aiming at the points whence the puffs of smoke came. But there was one daring fellow, Halleck by name, who climbed over the fence and amused himself shelling and eating the wheat while he dodged the bullets. So keen an eye did he keep out for the danger that he escaped without a scratch. While he was there, a man named Mattoon, a good soldier, came up, and seeing Halleck, jumped over with the exclamation, "What are you doing here?" "Just wait a minute and you will see," said Halleck. Mattoon was a fat, chubby fellow, and in just about a minute, a bullet struck him in the face, going through the fleshy part of the cheek and making the blood spout. "I told you so," said Halleck, who kept on eating wheat and defying the sharpshooters, who were unable to hit him, though he was a conspicuous target.

The secret of it was he did not stand still, but kept moving, and they had to hit him, if at all, like a bird on the wing which at the distance was a hard shot to make. The entire day was passed in this kind of skirmishing, and it was both dangerous and exciting. The men had lots of fun out of it, and only a few of them were shot, though there were many narrow escapes.

On the morning of July 14th, the Third Cavalry Division marched over the Hagerstown pike into Williamsport. There was no enemy there. Lee had given Meade the slip. His army was across the Potomac, in Virginia once more, safe from pursuit. As he reined up his faithful steed upon the northern bank of the broad river, the Union trooper looked wistfully at the country beyond. Well he knew that Lee had escaped like a bird from the snare, and could march leisurely back to his strongholds. Visions of the swamps of the Chickahominy, of Bull Run, of Fredericksburg, of Chancellorsville, passed before his mind as with pensive thought he gazed upon the shining valley of the Shenandoah, stretching away to the southward in mellow perspective. He wondered how long the two armies were to continue the work of alternately chasing each other back and forth across this battleground of the republic.

The wide, majestic river, no longer vexed by the splashing tread of passing squadrons, with smooth and tranquil flow swept serenely along, the liquid notes of its rippling eddies seeming to mock at the disappointment of the baffled pursuer. The calm serenity of the scene was in sharp contrast with the stormy passions of the men who sought to disturb it with the stern fatalities of war. The valley, rich with golden harvests, presented a charming dissolving view, melting away in the dim distance. On the left, the smoky summits of the Blue Mountains marked the eastern limits of this "storehouse of the Confederacy," the whole forming a picture in which beauty and grandeur were strikingly blended.

But this reverie of the soldier was soon rudely disturbed. Word came that they were not all across after all. Five miles below, at Falling Waters, in a bend of the river, was a ford where a portion of Longstreet's corps was yet to cross on a pontoon bridge. Kilpatrick started off in hot haste for Falling Waters, determined to strike the last blow on northern soil. The Sixth Michigan was in advance, two troops - B and F - under Major Weber, acting as advance guard. Kilpatrick and Custer followed Weber; then came Colonel Gray with the remainder of the regiment.

The march from Williamsport to Falling Waters was a wild ride. For the whole distance, the horses were spurred to a gallop.

Kilpatrick was afraid he would not get there in time to overtake the enemy, so he spared neither man nor beast. The road was soft and miry, and the horses sank almost to their knees in the sticky mud. For this reason the column straggled, and it was not possible to keep a single troop closed up in sets of fours. At such a rapid rate, the column plunged through the muddy roads, Weber and his little force leading.

On nearing Falling Waters, the column turned to the right through a wood which skirted a large cultivated field. To the right and front, beyond the field, was a high hill or knoll on which an earthwork had been thrown up. Behind the earthwork a considerable force of Confederate infantry was seen in bivouac, evidently taking a rest, with arms stacked. As a matter of fact, for it will be as well to know what was there, though the general in command made very little note of it at the time, there were two brigades - an entire division - commanded by General Pettigrew, one of the men who participated in Pickett's charge at Gettysburg.

On sighting this force, Custer ordered Weber to dismount his men, advance a line of skirmishers toward the hill and ascertain what he had to encounter. Kilpatrick however, ordered Weber to remount and charge the hill. At that time, no other portion of the regiment had arrived so as to support the charge.

Weber, knowing no law for a soldier except implicit obedience to orders, first saw his men well closed up, then placed himself at their head and giving the order "Forward," emerged from the woods into the open field, took the trot until near the top of the slope, close to the earthworks, and then with a shout the little band of less than a hundred men charged right into the midst of ten times their number of veteran troops. The first onset surprised and astonished the enemy, who had mistaken Weber's force for a squadron of their own cavalry. The audacity of the thing dazed them for a minute, and for a minute only.

Weber, cutting right and left with his saber and cheering on his men, pierced the first line, but there could be but one result.

Recovering from their surprise, the Confederate infantry rallied, and seizing their arms, made short work of their daring assailants. In a few minutes, of the three officers in the charge, two - Weber and Bolza - lay dead on the field, and the other - Crawford - had his leg shattered so it had to be amputated. The two brave troops were more than decimated, though a considerable number succeeded in escaping with their lives.

This charge which Kilpatrick in his official report characterized as "the most gallant ever made," was described by a Confederate eyewitness who was on the hill with Pettigrew and who wrote an account of the affair for a southern paper several years ago, as "a charge of dare-devils."

Major Peter Weber of the 6th Michigan Cavalry was killed leading a mounted charge at The Battle of Falling Waters. The attack also resulted in the death of Confederate General James Pettigrew.

In the meantime, just as Weber's command was repulsed, the other squadrons of the regiment began to arrive and were hurried across the field to the foot of the hill, and there dismounted to fight, dressing to the left as they successively reached the alignment and opening fire with their Spencers at once. But having disposed of the two mounted troops, the Confederates filled the earthworks and began to send a shower of bullets at those already formed or forming below.

My troop was the fourth from the rear of the regiment, and consequently several preceded it on the line. When I reached the fence along the side of the field next the woods, I found Lieutenant A.E. Tower, who since the death of Jewett had been acting adjutant, at the gap giving orders. He directed me to take my command across the field and form on the right of that next preceding. I had ridden so rapidly that only a few men had kept up the pace, and the remainder were strung out for some distance back. But taking those that were up, and asking the adjutant to tell the others to follow, I dashed into the field, and soon found that we were the targets for the enemy on the hill, who made the air vibrant with the whiz of bullets.

It was hot, but we made our way across without being hit, and reached the place where the regiment was trying to form under fire of musketry from the hill, and getting badly cut up. Reining up my horse, I gave the order, "Dismount, to fight on foot" and glancing back, saw my men coming in single file, reaching to the fence - probably an eighth of a mile - and the rear had not yet left the woods. The two leading sets of fours which alone were closed up obeyed the order, and dismounting to direct the alignment, I stepped in front of my horse, still holding the bridle rein in my right hand, when a minie bullet from the hill in front with a vicious thud went through my right foot, making what the surgeon in Washington afterwards said was the "prettiest wound I ever saw." I tried to stand but could not. The foot was useless. Private Halleck - the same who was eating wheat at Hagerstown a few days before - jumped to my rescue and helped me off the field.

Back of our position some distance, say 500 yards, was a log house in an orchard. To this we directed our steps, I leaning on Halleck's shoulder and hopping along on the unhurt foot. The most uncomfortable experience I had during the war I believe was during the passage across the open field to the orchard. Our backs were to the foe and the whistling bullets which came thick and fast all about served to accelerate our speed. I expected every moment to be shot in the back. One poor fellow, already wounded, who was trying to run to the rear, was making diagonally across the field from the right. As he was about to pass us, a bullet struck him and he fell dead in his tracks. Halleck succeeded in getting to the house, where he left me with the remark; "You are all right now, Captain; the boys need me and I will go back on the line." And back he went into the thickest of it, and fought gallantly to the end of the engagement, as I learned by inquiry afterwards.

After a little, the Confederates drove our line back beyond the house, and it was, for perhaps an hour, on the neutral ground between friends and foes. Shells from the opposing batteries hurtled around, and I did not know what moment one of them would come crashing through the building. A hospital flag had been displayed above it, which saved it.

Finally, sufficient force arrived to give our people the best of it, and the enemy was driven in confusion to the river, losing about 1,500 prisoners, one or two pieces of artillery and many small arms. General Pettigrew was killed by Weber or one of his men. Until the battle was over, I did not know what fearful losses had befallen the regiment. The total casualties were 33 killed and 56 wounded. The loss in officers was heavy: Major Weber, killed; Lieutenant Bolza, commanding troop B, killed; Lieutenant Potter, troop C, wounded and prisoner; Captain Royce, troop D, killed; Captain Kidd, troop E, wounded; Lieutenant Crawford, troop F, lost a leg; Lieutenant Kellogg, troop H, wounded and a prisoner.

The story of "The Michigan Cavalry Brigade in the Gettysburg Campaign," properly ends with the death of General Pettigrew and Major Weber at Falling Waters. No more brilliant passage at arms took place during the War for the Union, and it is a pity that some more able historian could not have written the story and immortalized the men, both dead and living, who had a part in it.[15]

FROM FALLING WATERS TO BUCKLAND MILLS

THE night following the Battle of Falling Waters, July 14, 1863, was a memorable one to the Michigan Cavalry Brigade, especially to those who like myself passed it in the field hospital. The log house into which the wounded were taken was filled with maimed and dying soldiers, dressed in Union blue. The entire medical staff of the division had its hands full caring for the sufferers. Many were brought in and subjected to surgical treatment only to die in the operation or soon thereafter. Probes were thrust into gaping wounds in search of the deadly missiles, or to trace the course of the injury. Bandages and lint were applied to stop the flow of blood. Splintered bones were removed and shattered limbs amputated. All night long, my ears were filled with the groans wrung from stout hearts by the agonies of pain, and the moans of the mortally hurt as their lives ebbed slowly away.

One poor fellow belonging to the First Michigan Cavalry was in the same room with me. He had a gunshot wound in the bowels. It was fatal and he knew it, for the surgeon had done his duty and told him the truth. He was a manly and robust young soldier who but a few hours before had been the picture of health, going into battle without a tremor and receiving his death wound like a hero. For hours, I watched and wondered at the fortitude with which he faced his fate. Not a murmur of complaint passed his lips. Racked with pain and conscious that but a few hours of life remained to him, he talked as placidly about his wound, his condition and his coming dissolution as though conversing about something of common, everyday concern. He was more solicitous about others than about himself, and passed away literally like one "who wraps the drapery of his couch about him and lies down to pleasant dreams." He died about three o'clock in the morning, and I could almost feel the reality of the flight of his tranquil spirit.

Field hospital. A forward station as close to the lines as possible served as the first stop for wounded men brought from the field. Each station would have basins, bandages, chloroform, opium and whiskey. Each medic had water, a tin cup, medicine and whiskey. The lightly wounded had to walk to the main field hospital, with those more seriously injured were treated here first.

In striking contrast to the picture thus presented was one in the room adjoining. Another trooper also fatally wounded, suffered so keenly from shock and pain that his fortitude gave way. He could not bear the thought of death. His nerve appeared to have deserted him and his anguish of mind and body, as he saw the relentless approach of the grim monster and felt his icy breath, will haunt my memory until I myself shall have joined the great army of Union veterans who are beyond the reach of pain and the need of pensions.

My own wound gave little annoyance except when the surgeon ran an iron called a probe into it, which attempt met with so vigorous a protest from his patient that he desisted, and that form of treatment stopped right there, so far as one cavalryman was concerned. The wound was well bandaged and plentiful applications of cold water kept out the inflammation.

Many of the officers and men came in to express their sympathy. Some of them entertained me with the usual mock congratulations on having won a "leave" and affected to regard me as a lucky fellow, while they were the real objects of sympathy. But the circumstances were such as to repress mirth or anything of that semblance. The regiment was in mourning for its bravest and best. The Sixth, having been the first regiment to get into the fight, had suffered more severely than any other. The losses had been grievous, and it seemed hard that so many bright lights of our little family should be so suddenly extinguished.

At daylight, I was still wide awake, but even amidst such scenes as I have described, fatigue finally overcame me and I sank into dreamland only to be startled, at first, by the fancied notes of the bugle sounding "to horse" or the shouts of horsemen engaged in the fray. At last however, "tired nature's sweet restorer" came to my relief and I fell into a dreamless sleep that lasted for several hours.

When I awoke, it was with a delightful sense of mind and body rested and restored. The wounded foot had ceased its pain. A gentle hand was bathing my face with cold water from the well, while another was straightening out the tangled locks, which to tell the truth were somewhat unkempt and overgrown from enforced neglect. Two ladies full of sympathy for the youthful soldier were thus kindly ministering to his comfort. As soon as fully awake to his surroundings, he opened his eyes and turned them with what was meant for a look of gratitude upon the fair friends who seemed like visiting angels in that place of misery and death.

It was an incongruous picture that presented itself - a strange blending of the gruesome sights of war with the beautiful environments of peace. The wonted tranquility of this rural household had been rudely disturbed by the sudden clangor of arms. A terrible storm of battle - the more terrible because unforeseen - had broken in upon the quietude of their home. In the early hours of the morning it had raged all around them. At the first sound of its approach, the terrified inmates fled to the cellar where they remained until it passed. They had come forth to find their house turned into a hospital. The kindness of those ladies is something that the Union trooper has never forgotten, for they flitted across his pathway, a transient vision of gentleness and mercy in that scene of carnage and suffering.

It was with a melancholy interest that I gazed upon the pallid face of my dead comrade of the First, who lay a peaceful smile upon his features which were bathed in a flood of golden light, as the hot rays of the July sun penetrated the apartment. The man in the hall was also dead. Others of the wounded were lying on their improvised couches, as comfortable as they could be made.

In the afternoon, the ambulance train arrived. The wounded were loaded therein and started for Hagerstown, bidding farewell to those who remained on duty, and who had already received marching orders which would take them back into "Old Virginia." The journey to Hagerstown was by way of Williamsport and the same pike we had marched over on the 6th of the month when Jewett was killed, and on the morning of the 14th when Weber was riding to one more saber charge at Falling Waters.

Nothing is more depressing than to pass over ground where a battle has recently been fought. Any veteran will say that he prefers the advance to the retreat - the front to the rear of an army. The true soldier would rather be on the skirmish line than in the hospital or among the trains. Men who can face the cannon's mouth without flinching, shrink from the surgeon's knife and the amputating table. The excitement, the noise, the bugle's note and beat of drum, the roar of artillery, the shriek of shell, the volley of musketry, the zip of bullet or ping of spent ball, the orderly movement of masses of men, the shouting of orders, the waving of battle flags - all these things inflame the imagination, stir the blood and stimulate men to heroic actions. Above all, the consciousness that the eyes of comrades are upon him, puts a man upon his mettle and upon his pride and compels him oftentimes to simulate a contempt for danger which he does not feel. The senses are too, in some sort, deadened to the hazards of the scene and in battle, one finds himself doing with resolute will things which under normal conditions would fill him with abhorrence.

Men fight from mingled motives. Pride, the fear of disgrace, ambition, the sense of duty - all contribute to keep the courage up to the sticking point. Few fight because they like it. The bravest are those who fully alive to the danger are possessed of that sublime moral heroism which sustains them in emergencies that daunt weaker men.

But when the excitement is over, when the pomp and circumstance are eliminated, when the unnatural ardor has subsided, when the tumult and rush have passed, leaving behind only the dismal effects - the ruin and desolation, the mangled corpses of the killed, the saddening spectacle of the dying, the sufferings of the wounded - the bravest would, if he could, blot these things from his sight and from his memory.

The night in the field hospital at Falling Waters did more to put out the fires of my military spirit and to quench my martial ambition than did all the experiences of Hunterstown and Gettysburg, of Boonsborough and Williamsport. And, as the ambulance train laden with wounded wound its tortuous way through the theater of many a bloody recent rencounter, it set in motion a train of reflections which were by no means pleasing. The abandoned arms and accouterments; the debris of broken down army wagons; the wrecks of caissons and gun carriages; the bloated carcasses of once proud and sleek cavalry chargers; the mounds showing where the earth had been hastily shoveled over the forms of late companions-in-arms; everything was suggestive of the desolation, nothing of the glory of war.

Ambulance train. The Rucker ambulance wagon was able to carry patients in seated or prone positions. Two stretchers on the floor of the wagon were divided by hinged joints and could be bent at right angles to serve as bench seats. The upper platform halves, hinged to the sides of the wagon, dropped down to form the seat backs. In the raised position, the seat backs were supported by iron posts hinged to their bottom surfaces. The resulting platform supported two more stretchers suspended from the roof.

It was nearly dark when the long train of ambulances halted in the streets of Hagerstown. Some large buildings had been taken for hospitals and the wounded were being placed therein as the ambulances successively arrived. This consumed much time, and while waiting for the forward wagons to be unloaded, it occurred to me that it would be a nice thing to obtain quarters in a private house. Barnhart, first sergeant of the troop, who accompanied me, proposed to make inquiry at once and ran up the stone steps of a comfortable looking brick house opposite the ambulance, and rang the bell. In a moment, the door opened and a pleasant voice inquired what was wanted. "A wounded officer in the ambulance yonder wants to know if you will take him in for a day or two until he can get ordered to Washington. He has funds to recompense you and does not like to go to the hospital." "Certainly," replied the voice; "bring him in." And Barnhart, taking me in his arms, carried me into the house, and guided to the second floor by the same lady who had met him at the door, deposited his burden on a couch in a well furnished apartment, and we were bidden to make ourselves at home.

In a little while, a nice hot supper of tea, toast, eggs and beefsteak, enough for both, was brought to the room by our hospitable hostess, who seemed to take the greatest pleasure in serving her guests with her own hands. Later in the evening, she called with her husband and they formally introduced themselves. They were young married people with one child, a beautiful little girl of six or eight summers. He was a merchant and kept a store in an adjoining building. They spent the evening in the room, chatting of the stirring events of the month, and indeed their experiences had been scarcely less exciting than our own.

Hagerstown had been right in the whirl of the battle storm which had been raging in Maryland. Both armies had passed through its streets and bivouacked in its environs. More than once, the opposing forces had contended for possession of the town. Twice the Union cavalry had charged in and driven the Confederates out, and once had been forced themselves to vacate in a hurry. It was almost inside its limits that Captain Snyder of the First Michigan Cavalry, serving on Kilpatrick's staff, had with the saber fought single handed five Confederate horsemen, and he was lying wounded mortally in a neighboring building. Our kind host and hostess entertained us until a late hour with interesting recitals of what they had seen from the inside or "between the lines."

That night after a refreshing bath, with head pillowed in down, I stowed myself away between snowy sheets for a dreamless sleep that lasted until the sun was high up in the eastern heavens. Barnhart was already astir and soon brought a surgeon to diagnose the case and decide what disposition should be made of the patient. Then the L--s and their little daughter came in with a cheery "good morning" and a steaming breakfast of coffee, cakes and other things fragrant enough and tempting enough to tickle the senses of an epicure. And, not content with providing the best of what the house afforded, Mr. L. brought in the choicest of cigars by the handful, insisting on my finding solace in the fumes of the fragrant weed. "Do not be afraid to smoke in your room," said the sunny Mrs. L., "my husband smokes and I am not the least bit afraid that it will harm the furnishings."

I glanced with a deprecatory gesture at the lace curtains and other rich furniture of the room, as much as to say, "Could not think of it," and in fact, before lighting a cigar, I took a seat by the open window where I sat and puffed the blue smoke into the bluer atmosphere, beguiling the time the while, talking with these good friends about the war. That was the very poetry of a soldier's life. For the better part of a week, the two cavalrymen were the guests of that hospitable family, who at the last, declined to receive any remuneration for their kindness.

The journey to Washington was by rail. In the cars, groups of interested citizens and soldiers as well questioned us eagerly for the latest news from the front, and our tongues were kept busy answering a steady fire of questions. No incident of the campaign was too trivial to find willing ears to listen when it was told. The operations of Kilpatrick's division seemed to be well known and there was much complimentary comment upon his energy and his dash. The name of Custer, "the boy general," was seemingly on every tongue, and there was no disposition on our part to conceal the fact that we had been with them.

Arriving at the capital in the middle of the day, we were driven to the Washington House, at the corner of Pennsylvania Avenue and 4 1/2 Street, where a room was engaged and preparations were made to remain until the surgeon would say it was safe to start for home.

The Washington House

The Washington House was a hotel of the second class, but many nice people stopped there. Among the regular guests was Senator Henry Wilson of Massachusetts, afterwards elected vice-president on the ticket with Grant. He was a very modest man, plain in dress and unassuming in manner. No one would have suspected from his bearing that he was a senator and from the great commonwealth of Massachusetts. The colleague of Charles Sumner, Henry Wilson was at that time one of the ablest, most widely known and influential statesmen of his day. Conspicuous among the anti-slavery leaders of New England, his voice always had been heard in defense of human rights. His loyalty to the Union was equaled only by his devotion to the interests of the soldiers. He lived a quiet, unostentatious life at the hotel, where his well-known face and figure could be seen when the Senate was not in session. He was a man of strong mentality, of sturdy frame and marked individuality.

As chairman of the Committee on Military Affairs, he had been able to make himself extremely useful to the government in the prosecution of the war, and the soldiers found in him always a friend. He was very agreeable and companionable, and did not hold himself aloof from the common herd, as smaller men in his position might have done. He was seen often chatting with other guests of the house when they were gathered in the parlors, after or awaiting meals. Once, I met him at an impromptu dancing party, and he entered into the amusement with the zest of youth.

A month in Washington, and a surgeon's certificate secured the necessary leave, when accompanied by Lieutenant C.E. Storrs of troop B, who had been severely wounded in one of the engagements in Virginia after Falling Waters, I started by the Pennsylvania Line for the old home in Michigan, stopping a couple of days en route at Altoona to breathe the fresh mountain air.

Senator Henry Wilson. Before and during the Civil War, he was a strong opponent of slavery. After the Civil War, he supported the "Radical" program for Reconstruction. He was elected Vice President as running mate with Ulysses S. Grant, and served from March 4, 1873 until his death on November 22, 1875.

Resuming the journey, we reached Pittsburg, to be met at the station by a committee stationed there for the purpose of looking out for the comfort of all soldiers who passed through the city, either going or coming. We were conducted to a commodious dining hall where a free dinner, cooked and served by the fair hands of the patriotic ladies of the "Smoky City," was furnished. It was an experience which left in our minds a most grateful appreciation of the noble spirit that actuated the Northern women in war times.

It was scarce two-thirds of a year since, as schoolboys innocent of war, though wearing the Union blue, we had gone forth to try our mettle as soldiers, and it needs not to be said that there was a warm welcome home for the veterans fresh from one of the most memorable military campaigns in all the history of the world. The greetings then and there received were ample compensation for all that we had done and dared and suffered. I can never forget how kind the people were; how they gathered at the railroad station; how cordially they grasped us by the hand; how solicitous they were for our comfort; how tenderly we were nursed back to health and strength; how fondly an affectionate mother hung upon every word as we told the story of the exploits of the boys in the field; how generously the neighbors dropped in to offer congratulations; how eagerly they inquired about absent friends; how earnestly they discussed the prospect of ultimate victory; how deep and abiding was their faith in the justice of the cause and in the ability of the government to maintain the Union; and how determined that nothing must be held back that was needed to accomplish that result. For some days, there was a regular levee beneath my father's roof, and the good people of the town gave the Union soldier much cause to remember them with gratitude as long as he lives.

Only in a single instance was anything said that seemed obnoxious to a nice sense of propriety, or that marred the harmony of an almost universally expressed sentiment of patriotic approval of what was doing to preserve the life of the nation - a sentiment in which partisanism or party politics cut no figure whatever. One caller had the bad taste to indulge in severe and unfriendly criticism of "Old Abe," as he called the president. That was going too far, and I defended Mr. Lincoln against his animadversions with all the warmth, if not the eloquence, of the experienced advocate - certainly with the earnestness born of a sincere admiration for Abraham Lincoln and love of his noble traits of character, his single hearted devotion to his country.

I had seen him in Washington weighed down with a tremendous load of responsibility such as few men could have endured. I had noted as I grasped his hand the terrible strain under which he seemed to be suffering; the appearance of weariness which he brought with him to the interview; the pale, anxious cast of his countenance; the piteous, far-away look of his eyes; and by all these tokens he said, as plainly as if he had put it into words; "Love and solicitude for my country are slowly, but surely, wearing away my life." I saw shining through his homely features the spirit of one of the grandest, noblest, most lovable of the characters who have been brought by the exigencies of fate to the head of human affairs.

The soldiers loved him and they idealized him. He was to them the personification of the Union cause. The day for the discussion of abstract principles had long gone by. Their ideal had ceased to be an impersonal one. All the hope, the faith, the patriotism of the soldiers centered around the personality of the president. In their eyes and thoughts, he stood for the idea of nationality as Luther stood for religious liberty, Cromwell for parliamentary privilege, or Washington for colonial independence. To blame him was to censure the boys in blue and the cause for which they fought. No man whose heart was not wholly with the Northern armies in the struggle could rise to an appreciation of the character of Lincoln.

But the great heart of the North never ceased to beat in

harmony with the music of the Union. The exceptions to the rule were so rare as to scarcely merit notice. The "copperheads" and "knights of the golden circle" will hardly cut so much of a figure in history as do the Tories of the Revolution.

On the 11th day of October, 1863, after an absence of three months duration, during which time I had been commissioned major to fill the vacancy caused by the death of Weber, I took passage at Washington on a ramshackle train over the Orange and Alexandria railroad to go to the front again. Storrs, whose wound had healed, joined me and we made the journey together.

The train reached Bealton Station, north of the Rappahannock River, a little before dark. The harbingers of a retreating army were beginning to troop in from the front. The Army of the Potomac was falling back toward the fastnesses of Centerville, the Army of Northern Virginia in close pursuit. Meade, who in July was chasing Lee across the Potomac back into Virginia, was himself now being hurried by Lee over the Rappahannock. The tables had been completely turned. The pursued had become the pursuer.

As usual, the flanking process had been resorted to. Using his cavalry as a screen, Lee was attempting to maneuver his infantry around Meade's right, and after the manner of Stonewall Jackson in the Second Bull Run campaign of 1862, interpose between the Federal army and Washington. Thanks to the vigilance of his outposts, the Union commander detected the movement in time and was able to thwart the strategy of his able adversary. Keeping his army well in hand, he retreated to Bull Run, Fairfax and Centerville.

While this was going on, there was a series of spirited encounters between the Union and Confederate cavalry, commanded by Pleasonton and Stuart respectively - the former bringing up the rear and covering the retreat, the latter bold and aggressive as was his wont. These affairs, which began on the 9th, culminated on the 11th in one of the most exciting, if not brilliant, engagements of the war, Kilpatrick taking a prominent part, second only to that performed by the heroic John Buford and his First Cavalry Division.

When the movement began on the evening of the 9th, Fitzhugh Lee was left to hold the line along the south bank of the Rapidan River, Buford's cavalry division confronting him on the north side. Stuart, with Hampton's division of three brigades, Hampton being still disabled from the wounds received at Gettysburg, spent the 10th swarming on the right flank of the Confederate army in the country between Madison Court House and Woodville on the Sperryville pike. Kilpatrick was in the vicinity of Culpeper Court House. Stuart succeeded not only in veiling the movements of the Confederate army completely, but on the morning of the 11th found time to concentrate his forces and attack Kilpatrick at Culpeper. Buford crossed the Rapidan to make a reconnoissance, and encountering Fitzhugh Lee, recrossed at Raccoon Ford, closely followed by the latter. The pursuit was kept up through Stevensburg, Buford retreating toward Brandy Station.

When Stuart heard Fitzhugh Lee's guns, he withdrew from Kilpatrick's front and started across country, intending to head off the federal cavalry and reach Fleetwood, the high ground near the Brandy Station in advance of both Buford and Kilpatrick. The latter however, soon discovered what Stuart was trying to do and then began a horse race of three converging columns toward Brandy Station, Stuart on the left, Buford followed by Fitzhugh Lee on the right, and Kilpatrick in the center.

Buford was in first and took possession of Fleetwood. Rosser with one of Lee's brigades, formed facing Buford, so that when the head of Kilpatrick's column approached, Rosser was across its path, but fronting in the direction opposite to that from which it was coming. Kilpatrick, beset on both flanks and in rear, and seeing a force of the enemy in front also, and ignorant of Buford's whereabouts, formed his leading regiments and proceeded to charge through to where Buford was getting into position. This charge was led by Pleasonton, Custer and Kilpatrick in person. Rosser, seeing what was coming and caught between two fires, dexterously withdrew to one side, and when the rear of Kilpatrick's division was opposite to him, charged it on one flank while Stuart assaulted it on the other, and there was a general melee in which each side performed prodigies of valor and inflicted severe damage on the other. The First and Fifth Michigan regiments were with the advance while the Sixth and Seventh helped to bring up the rear.

Brandy Station. The Battle of Brandy Station was the largest predominantly cavalry engagement of the Civil War, and the largest ever to take place on American soil.

The rear of the column had the worst of it and was very roughly handled. The two divisions having united, Pleasonton took command, and bringing his artillery hurriedly into position, soon had Stuart whipped to a standstill. All the fighting in this battle was done on horseback and no more daring work was done by either side, on any of the battlefields of the war, than was seen at Brandy Station. Those who were in it describe it as the most stirring and picturesque scene that they ever witnessed; especially when the three long columns, one of blue and two of gray, were racing on converging lines toward the objective point on Fleetwood Hill.

It must have been a pretty picture: Buford hurrying into line to face to the rear; the Federal batteries unlimbering and going into position to resist the coming attack; Rosser galloping front into line to find himself attacked front and rear; Kilpatrick, with Rosser in his front, Fitzhugh Lee and Stuart on his flanks; detachments breaking out of the Confederate columns to attack the flanks and rear of Kilpatrick's flying division; Federal regiments halting and facing toward the points of the compass whence these attacks came; then falling back to new positions, stubbornly contesting every inch of ground; the fluttering of guidons and battle flags, the flash of sabers and puffs of pistol shots - altogether a most brilliant spectacle.   Stuart was kept at bay until after nightfall, when Pleasonton withdrew in safety across the river.

It has been claimed that Brandy Station was the greatest cavalry engagement of the war.   Sheridan, who was then still in the west, and consequently not there, awards that honor to Yellow Tavern, fought the following season.   Doubtless he was right, for the latter was a well planned battle in which all the movements were controlled by a single will.   But most of the fighting at Yellow Tavern was done on foot, though Custer's mounted charge at the critical moment won the day.

Brandy Station was a battle in which all the troopers were kept in the saddle. It was however, a battle with no plan; though it is conceded that Pleasonton handled his command with much skill after the two divisions had united.   His artillery was particularly effective. Captain Don G. Lovell of the Sixth Michigan, the senior officer present with the regiment, greatly distinguished himself in the difficult duty of guarding the rear, meeting emergencies as they arose with the characteristic courage and coolness which distinguished him on all occasions on the field of battle. The battle ended about the time our train reached Bealton, so Storrs and I missed the opportunity of taking part in one of the most memorable contests of the Civil War.

After a night on the platform of the railroad station, we started at dawn to find the brigade. From wounded stragglers, the salient events of the previous day were learned and the inference drawn from the information which they were able to give was that the cavalry must be encamped somewhere not far away. All agreed that it was having a lively experience. Everything however, was at sixes and sevens and it was only after a long and toilsome search that the regimental quartermaster was located among the trains. My horse, equipments and arms had disappeared, but fortunately Storrs found his outfit intact, and having two mounts, he loaned me one. Selecting from the quartermaster's surplus supplies a government saber, revolver and belt, thus equipped and mounted on Storrs's horse, I rode in search of the regiment, which we ascertained to be in camp in the woods, some distance away from the trains.

Brandy Station Railroad Depot

When at last found, it proved to be a sorry looking regiment, but a wreck and remnant of its former self. With two troops (I and M) absent on detached service in the Shenandoah Valley, the

Sixth Michigan started in the Gettysburg campaign, June 21st, with between 500 and 600 troopers in the saddle. When Storrs and I rode into that silvan camp on that bright October morning, there were less than 100 men present for duty, including not a single field officer. Many of the troops were commanded by lieutenants, some of them by sergeants, and one had neither officer nor non-commissioned officer.

Libby Prison. In 1889, the building was purchased by Charles F. Gunther, a candy-maker. It was disassembled and moved to Chicago. There, it was rebuilt and renovated to serve as a war museum. After the museum failed to draw crowds, the building was dismantled in 1895 and sold in pieces as souvenirs.

They had been fighting, marching and countermarching for months, and had a jaded, dejected appearance, not pleasant to look upon and very far removed indeed from the buoyant and hopeful air with which they entered upon the campaign. At one point during the retreat of the day before, it had been necessary to leap the horses over a difficult ditch. Many of them fell into it, and the riders were overtaken by the enemy's horse before they could be extricated. Among these was Hobart, sergeant major, who was taken to Libby prison where he remained until the next year, when he was exchanged.

The next thing was to report to General Custer for duty. It was my first personal interview with the great cavalryman. He was at his headquarters in the woods, taking life in as light hearted a way as though he had not just come out of a fight, and did not expect others to come right along. He acted like a man who made a business of his profession; who went about the work of fighting battles and winning victories, as a railroad superintendent goes about the business of running trains. When in action, his whole mind was concentrated on the duty and responsibility of the moment; in camp, he was genial and companionable, blithe as a boy. Indeed he was a boy in years, though a man in courage and in discretion.

George A. Custer (seated, right) and officers.

After drawing rations and forage, the march was resumed, and little of incident that was important intervening, on the 14th the division was encamped on the north side of Bull Run near the Gainesville or Warrenton turnpike, where we remained undisturbed until the evening of the 18th when the forward movement began which culminated on the 19th in the Battle of Buckland Mills, which will be the theme of the next chapter.

THE BATTLE OF BUCKLAND MILLS

BUCKLAND MILLS was, in some sort, a sequel to Brandy Station. The latter battle was a brilliant passage at arms, in which neither side obtained a decisive advantage. Kilpatrick was still pugnacious and both willing and anxious to meet Stuart again. That his mind was full of the subject was evinced by a remark he was heard to make one morning at his headquarters on the Bull Run battleground. He was quartered in a house, his host a Virginian too old to be in the army, and who remained at home to look after the property. It was a clear day, and when the general came out on the porch, the old gentleman accosted him with a cheery; "A fine day, general!" "Yes, a fine day for a fight;" was the instant reply.

In most men, this would have sounded like gasconade. In Kilpatrick's case, it was not so considered. He was credited with plenty of pluck, and it was well understood that he was no sooner out of one action than he was planning to get into another. He ran into one a day or two later, which furnished him all the entertainment of that kind that he wanted, and more too.

Reconnoissances across Bull Run on the Gainesville Road disclosed a considerable force of mounted Confederates. When their pickets were driven in by the Sixth Michigan on the 15th and again by the First Michigan on the 16th, strong reserves were revealed. As a matter of fact, Stuart was at Buckland Mills with Hampton's division, and Fitzhugh Lee was at or near Auburn but a few miles away. They had their heads together and devised a trap for Kilpatrick, into which he rode with his eyes shut.

Sunday evening, October 18th, the Third Division moved out across Bull Run, Kilpatrick in command, Custer's brigade leading, Davies[16] with the First Brigade bringing up the rear. Stuart's cavalry was attacked and driven rapidly until dark by the First Vermont Cavalry[17] under Lieutenant Colonel Addison W. Preston, acting as advance guard. Early on Monday morning, October 19th, the march was resumed, the Sixth Michigan in advance.

About midway between Bull Run and Broad Run, the Confederate rear guard, a regiment of Young's brigade of Hampton's division was encountered, which fell back before the advance of the Sixth Michigan, making but slight resistance and retreating across Broad Run, where it was found that Stuart had taken up a strong position, forming the three brigades of Gordon, Rosser and Young in line on the opposite side, as if to contest the crossing.

The stream was deep and difficult, spanned at the pike by a stone bridge. Its banks were wooded. Stuart stationed a piece of artillery on the high ground so as to command the bridge and its approaches. A portion of the regiment was dismounted and advanced to engage the dismounted Confederates across the stream. Captain George R. Maxwell of the First Michigan, whose regiment was at the time in the rear, rode up and asked permission to take a carbine and go on foot with the men of the Sixth who were in front. The permission was granted, and giving his horse into the charge of an orderly, he was in a few moments justifying his already well established reputation as a man of courage by fighting like an enlisted man on the skirmish line of a regiment not his own, thus voluntarily exceeding any requirements of duty.

Custer rode up with his staff and escort and halted in the road, making a conspicuous group. Stuart's cannoneers planted a shell right in their midst, which caused a lively scattering, as they had no desire to be made targets of for that kind of artillery practice. Fortunately, no one was killed.

Custer then brought up his entire command and formed a line of battle, the Sixth Michigan in the center across the pike, the Fifth Michigan on the right, the Seventh Michigan on the left, the First Michigan and First Vermont in reserve, mounted. After a somewhat stubborn resistance, Stuart apparently reluctantly withdrew, permitting Custer to cross, though he could have held the position easily against ten times his number, whereas as the sequel proved, he greatly outnumbered Kilpatrick. The Seventh crossed at a ford about a mile below, the other regiments at the bridge. Stuart retreated toward Warrenton.

It was then about noon, perhaps a little later than that. Kilpatrick came up and ordered Custer to draw in his skirmishers and allow Davies to pass him and take the advance. Custer massed his command on some level ground behind a hill, beyond the bridge and adjacent to the stream. Davies crossed the bridge, passed the Michigan Brigade and took up the pursuit of Stuart. Kilpatrick, with his staff, followed along the pike in rear of Davies's brigade. As he was moving off, Kilpatrick directed Custer to follow the First Brigade and bring up the rear.

This was the very thing that Stuart was waiting for. It had been arranged between him and Fitzhugh Lee that he, with his three brigades,[18] was to fall back without resistance before the two brigades of the Third Division, until they were drawn well away from the bridge, when Lee, who was coming up from Auburn through the woods to the left with the brigades of Lomax, Chambliss and Wickham and Breathed's battery, would swing in across the pike, cut Kilpatrick off from the bridge, and then, at the first sound of Lee's guns, Kilpatrick was to be attacked simultaneously by Stuart in front and by Lee in rear and thoroughly whipped.

It was a very pretty bit of strategy and came very near being successful. The plan was neatly frustrated by one of those apparent accidents of war which make or unmake men, according as they are favorable or unfavorable. Custer respectfully but firmly demurred to moving until his men could have their breakfast - rather their dinner, for the forenoon was already spent. Neither men nor horses had had anything to eat since the night before, and he urged that the horses should have a feed and the men have an opportunity to make coffee before they were required to go farther.

Custer was a fighting man, through and through, but wary and wily as brave. There was in him an indescribable something - call it caution, call it sagacity, call it the real military instinct - it may have been genius - by whatever name entitled, it nearly always impelled him to do intuitively the right thing. In this case, it seemed obstinacy, if not insubordination. It was characteristic of him to care studiously for the comfort of his men. And he did not believe in wasting their lives. It is more than probable that there was in his mind a suspicion of the true state of things. If so, he did not say so, even to the general commanding the division. He kept his own counsel and had his way. The delay was finally sanctioned by Kilpatrick, and the brigade remained on the bank feeding their horses and making coffee, Davies meanwhile advancing cautiously on the Warrenton road to a point within about two or three miles of Warrenton. Stuart made slight if any attempt to resist his progress.

The Gainesville-Warrenton pike, after crossing Broad Run, is bounded on both sides by cleared farm lands, fringed about one third of a mile back by woods. From the place of Custer's halt, it was not more than 500 or 600 yards to these woods. The road runs in a westerly direction and the brigade was on the south side of it.

There is very little of record from which to determine the time consumed by Custer's halt. It is a peculiar circumstance that not a single report of this battle made by a regimental commander in Custer's brigade appears in the official war records. A similar omission has been noted in the Battle of Gettysburg. Custer made a report and so did Kilpatrick and Davies, but they are all deficient in details. There is no hint in any of them as to the duration of the delay. The Confederate chronicles are much more complete. From them, it would appear that the stop was made about noon and that the real battle began at 3:30 in the afternoon. Memory is at fault on this point for the reason that after coffee and while the horses were feeding, I lay down upon the ground and fell asleep. Before that, some of the men had gone into the adjacent fields in search of long forage. It was understood that the Seventh Michigan, after crossing at the lower ford, was scouting through the country toward Greenwich and there was no hint or suspicion that an enemy could approach from that direction without being discovered by this scouting party.

Finally Custer was ready to move. Awakened by a staff officer, I was directed to report to the general. "Major," said he, "take position with your regiment about 500 yards toward those woods remain there until the command is in column on the pike, then follow and bring up the rear." The order was given with a caution to be careful, as the Seventh Michigan had been scouting near Greenwich and might be expected to come in from that direction. Greenwich is almost due south from Buckland Mills, whereas Auburn, from which place Fitzhugh Lee was approaching, lay considerably west of south.

The movement of the two commands began simultaneously. The Fifth Michigan, Pennington's battery, the First Michigan and First Vermont, with Custer and his staff leading, were in a few moments marching briskly in column on the Warrenton pike, which was not very far away from the starting point. The Sixth Michigan meantime proceeded in column of fours toward the place designated by General Custer, close up to the woods. Nothing had been seen or heard of Davies for some time. Everything was quiet. Nothing could be heard except the tramp of the horses' feet and the rumble of the wheels of Pennington's gun carriages, growing more and more indistinct as the distance increased.

Warrenton Pike. The stone house sits on a hill above the Bull Run battlefield.

The Sixth had gone about 250 or 300 yards and was approaching a fence which divided the farm into fields when Captain Don G. Lovell, who was riding by the side of the commanding officer of the regiment,[19]suddenly cried out; "Major, there is a mounted man in the edge of the woods yonder," at the same time pointing to a place directly in front and about 200 yards beyond the fence.

Captain Lovell was one of the most dashing and intrepid officers in the brigade. He was always cool and never carried away with excitement under any circumstances. It is perhaps doubtful whether he could have maintained his customary imperturbability if he had realized, at the moment, just what that lone picket portended.

A glance in the direction indicated revealed the truth of Captain Lovell's declaration, but recalling what General Custer had said, I replied; "The general said we might expect some mounted men of the Seventh from that direction." "But that vidette is a rebel," retorted Lovell, "he is dressed in gray." "It can't be possible," was the insistent reply, and the column kept on moving. Just then, the man in the woods began to ride his horse in a circle. "Look at that," said Lovell; "that is a rebel signal; our men don't do that."

The truth of the inference was too evident to be disputed. Things were beginning to look suspicious and in another instant, all doubt, if any remained, was set at rest. The horseman, after circling about a time or two, brought his horse to a standstill facing in the direction from which we were approaching. There was a puff of smoke from the muzzle of his revolver or carbine, and a bullet whizzed by and buried itself in the breast of one of the horses in the first set of fours. "There,--it," exclaimed Lovell. "Now you know it is a rebel, don't you?"

The information was too reliable not to be convincing, and the regiment was promptly brought front into line, which had hardly been accomplished when shots began to come from other points in the woods, and no further demonstration was needed that they were full of Confederates.

The fence was close at hand, and the command to dismount to fight on foot was given. The Sixth deployed along the fence and the Spencers began to bark. The horses were sent back a short distance under cover of a reverse slope. The acting adjutant was dispatched to overtake Custer and report to him that we were confronted by a large force of Confederates and had been attacked. Before he had started, the Confederates displayed a line of dismounted skirmishers that extended far beyond both flanks of the regiment and a swarm of them in front. A Michigan regiment, behind a fence and armed with Spencer carbines, was a dangerous antagonist to grapple with by a direct front assault, and Fitzhugh Lee's men were not eager to advance across the open field, but hugged the woods, waiting for their friends on the right and left to get around our flanks, which there was imminent danger of their doing before relief could come. It did not however, take Custer long to act.

Putting the Fifth Michigan in on the right of the Sixth, he brought back Pennington's battery and stationed the First Vermont mounted to protect the left flank, holding the First Michigan mounted in reserve to support the battery and to reinforce any weak point, and proceeded to put up one of the gamiest fights against odds seen in the war. Opposed to Custer's five regiments and one battery, Fitzhugh Lee had twelve regiments of cavalry, three brigades under Lomax, Owen and Chambliss and as good a battery – Breathed's - as was in the Confederate service.

Before the dispositions described in the foregoing had been completed, Breathed's battery, which had been masked in the woods to the right and front of the position occupied by the Sixth Michigan, opened fire with shell. But Pennington came into position with a rush, and unlimbering two pieces in less time than it takes to tell it, silenced the Confederate artillery, firing over the heads of the Sixth Michigan skirmishers.

Fitzhugh Lee pressed forward his dismounted line, following it closely with mounted cavalry, and made a desperate effort to cut off Custer's line of retreat by the bridge. This he was unable to do. The Sixth held on to the fence until the Confederates were almost to it, and until ordered by Custer to retire, when they fell back slowly, and mounting their horses, crossed the bridge leisurely without hurry or flurry, the battery and the other regiments, except the First and Fifth Michigan, preceding it. The First Michigan brought up the rear.

Fitzhugh Lee was completely foiled in his effort to get in Custer's rear or to break up his flanks. Unfortunately, a portion of one battalion of the Fifth Michigan, about fifty men, under command of Major John Clark, with Captain Lee and Adjutant George Barse, was captured. Being dismounted in the woods on the right, they were not able to reach their horses before being intercepted by the enemy's mounted men.

Custer, on the whole, was very fortunate and had reason to congratulate himself on escaping with so little damage. Davies did not fare so well. When Kilpatrick found that Custer was attacked, he sent orders to Davies to retreat. But the sound of firing which gave this notice to Kilpatrick was also the prearranged signal for Stuart, and that officer immediately turned on Davies with his entire division, and Davies, though he put up a stout resistance, had no alternative finally but to take to the woods on the north side of the pike and escape, every man for himself. Fitzhugh Lee was between him and the bridge, he was hemmed in on three sides, and in order to escape, his men had to plunge in and swim their horses across Broad Run. The Fifth Michigan, except Major Clark's command, escaped in the same way. The wagons which followed Davies, including Custer's headquarters wagon containing all his papers, were captured.

At first blush, it may appear that, if the vidette who fired the first shot, thus divulging the fact of the enemy's presence, had not done so, the Sixth Michigan would have gone on and marched right into Fitzhugh Lee's arms. It is not likely however, that such would have been the result. Captain Lovell had already seen and called attention to the picket, declaring that he was a rebel. The obvious course, under the circumstances, before taking down the fence and advancing to the woods, would have been to deploy a skirmish line and feel of the woods instead of blundering blindly into them.

Fitzhugh Lee made a mistake in halting to dismount. He should have charged the Sixth Michigan. Had he charged at once, mounted as Rosser did in the Wilderness, with his overwhelmingly superior force at the moment of his arrival, he must certainly have interposed between Custer and the bridge. He allowed one regiment to detain his division until Custer could bring back his brigade and get his regiments into position to support each other.

Major H.B. McClellan, Stuart's adjutant general, commenting

in his book[20] on this battle, says that "Custer was a hard fighter, even on a retreat." He also says "Fitzhugh Lee had come up from Auburn expecting to gain, unopposed, the rear of Kilpatrick's division, but he found Custer's brigade at Broad Run ready to oppose him. A fierce fight ensued."

Major McClellan also quotes Major P.P. Johnston, who commanded a section of Breathed's battery in the fight, as saying "My battery was hotly engaged. The battle was of the most obstinate character, Fitz Lee exerting himself to the utmost to push the enemy, and Custer seeming to have no thought of retiring."

The battle was opened by Wickham's brigade of Virginians commanded by Colonel T.H. Owen of the Third Virginia Cavalry. It was the First, Second and Third Virginia that led the advance. Pennington gave Breathed's battery much the worst of it.

The truth is that Fitz Lee did not find Custer ready to oppose him, though it did not take him long to get ready after he was attacked. Custer with most of his command was well on his way to follow Kilpatrick. Only one regiment was left behind, and that one regiment - the Sixth Michigan cavalry - was taken entirely by surprise when fired upon by the vidette, and was all that Colonel Owen had in front of him when he arrived and began the attack.

It is possible that ignorance of what it was facing helped the Sixth Michigan to hold on until Custer could be notified and brought back. And again, it is possible that Custer was marching more slowly than the writer knows of; that he suspected the ruse which was being played by his old West Point instructor,[21] and sent the regiment out there for the express purpose of developing the enemy, if enemy there was, making a feint of moving away so as to deceive, but keeping an ear to windward to catch the first sound of danger.

It has always seemed to the writer that General Custer must have had a motive which did not appear on the surface in giving that order. His order was to go 500 yards. Five hundred yards would have brought us to the woods. If he suspected that there might be an enemy there, no surer way to find out whether his suspicions were well founded or not could have been chosen. One thing is certain. He was back in an incredibly short space of time. It may be that he heard the sound of firing and was on his way when the adjutant found him.

Fitzhugh Lee. The nephew of Robert E. Lee, Fitzhugh Lee commanded the cavalry of the Army of Northern Virginia during the last months of the Civil War. At Appomattox Court House in 1865, he refused to give up. He slipped around the flanks of the Union Army and made his way to Lynchburg, Virginia. Three days later, realizing the futility of continued resistance, he returned to Appomattox and surrendered.

Fitzhugh Lee followed Custer half way to Gainesville and then withdrew. Near that place was found a line of Federal infantry sent out to support the cavalry, but it did not advance far enough to get into the fight. That night, Kilpatrick invited all the officers of the division to his headquarters and made a sorry attempt at merry making over the events of the day. There were milk punch and music, both of very good quality, but the punch, palatable as it undeniably was, did not serve to take away the bad taste left by the affair, especially among the officers of the First Brigade. Custer's men did not feel so badly. They had saved their bacon and their battery, and the wariness, prudence and pluck of their young commander had prevented a much more serious disaster than had actually happened.

It may be of interest enough to mention that Fitz Lee told the writer, in Yorktown, in 1881, that Stuart was at fault in stopping to fight at Buckland Mills; that under the arrangement with him (Lee), Stuart should have fallen back very rapidly without making any resistance whatever, until he had lured Kilpatrick with his entire division some distance beyond the bridge. In that event, General Lee would have found the opportunity he was seeking. But he did not know about Custer's action in insisting on stopping there. He was much surprised when informed of the true state of things, since he had felt that Stuart was blameworthy in the matter. He had supposed that it was Stuart's resistance to the Federal advance which kept Custer's brigade back until his arrival, and foiled his well planned attempt.

WINTER QUARTERS IN STEVENSBURG

IN the month of November, 1863, the Army of the Potomac recrossed the Rappahannock and the Army of Northern Virginia retired behind the Rapidan. General Meade took up the line through Culpeper, placing the Third Division on the left flank with headquarters at Stevensburg. The advance into Stevensburg was stoutly contested by Hampton's division, and the Confederate cavalry showed that it had not lost any of its fighting qualities, if its dash and spirit had been somewhat dampened by the sturdy resistance put up in the recent campaign by the Federal troopers led by Pleasonton, Buford, Gregg, Kilpatrick and Custer.

At the time of the Mine Run affair, the Michigan cavalry crossed the Rapidan at Morton's Ford and attacked Ewell's infantry, falling back after dark to the old position on the north side of the river.

After that episode, the army went into winter quarters. The three generals - Kilpatrick, Custer and Davies - had quarters in houses, the rest for the most part lived in tents or huts. The Sixth was hutted in temporary structures built of logs surmounted by tents. They were fitted with doors, chimneys and fireplaces - some of them with sashes and glass and were very comfortable. The winter was a very cold one. There was some snow, even in Virginia, and the first day of January, 1864 is still remembered as noteworthy for its extremely low temperature throughout the country.

While in this camp, the Michigan regiments had a visit from Jacob M. Howard, the colleague of Zachariah Chandler in the United States Senate. He was one of the ablest men who ever represented the state in the national Congress. He had served with high distinction as Attorney General of the state before being elected to the Senate. As chairman of the Senate Committee on Pacific Railroads, he had much to do with piloting the country through the many difficulties which stood in the way of the accomplishment of the great enterprise of laying tracks for the iron horse across the American desert - spanning the continent with railroads - and reducing the journey from the Missouri River to the Pacific Ocean from one of months to one of days - the most important of the achievements that followed close on the heels of the Civil War. The senator made a patriotic speech to the soldiers and was cordially cheered.

Winter camp. Winter huts were typically built by the armies from the surrounding materials such as trees, mud, leaves and canvases.

The cavalry picket line was twenty-five miles long, and it was no child's play to serve as field officer of the day, when every picket post and every vidette had to be visited at least once each twenty-four hours. The outer line was along the Rapidan River. The Confederate pickets on the other side were infantry. The Union pickets were mounted and the duty was very wearing on both men and horses. Stuart's cavalry performed comparatively but little picket duty, and was kept back in comfortable quarters, recruiting and fitting for the coming spring campaign.

Jacob Howard was a U.S. Representative and U.S. Senator from the state of Michigan during and after the American Civil War. He worked closely with Abraham Lincoln in drafting and passing the Thirteenth Amendment, which abolished slavery. In the Senate, he also served on the Joint Committee on Reconstruction.

During the winter, there was very little firing between the pickets. There was a sort of tacit understanding that they were not to molest each other. Indeed, officers could ride along the line without fear of being shot at. When on inspection duty, they at times rode down to the bank and conversed with the enemy on the other side. The pickets were suspected of crossing and recrossing and exchanging civilities – trading tobacco for papers and the like. The word of honor would be given to allow the Federal or Confederate, as the case might be, to return in safety, and it was never violated when given. These visits were always in the daytime of course, for at night vigilance was never relaxed, and a vidette was not supposed to know anybody or permit even his own officers to approach without the proper countersign.

Thorton Stringfellow was the pastor of Stevensburg Baptist Church in Culpeper County, Virginia. He is best known for his 1856 book, *Scriptural and Statistical Views in Favor of Slavery.*

Life in winter quarters was at best dull, and it relieved the monotony to go on picket. The detail as field officer of the day was welcomed, although it necessitated a ride of forty or fifty miles and continuous activity for the entire of the tour of duty, both night and day. On these rides, I made the acquaintance of a number of Virginia families who lived near the river and within our lines. Of these, I can now recall but two. On the banks of the Rapidan, directly in front of Stevensburg, lived a man named Stringfellow, who owned a large plantation which had been despoiled of everything of value except the house and a few outbuildings. Every fence was gone, and not a spear of anything had been permitted to grow. Mr. Stringfellow was a tall man with gray hair, and clerical in garb and aspect. He was, in fact, a clergyman, and the degree of doctor of divinity had been conferred upon him - a thing that in those days meant something. Degrees, like brevets, were not so easily obtained before the Civil War period as they have been since.

Mr. Stringfellow was a gentleman of culture, a scholar and profound student of Biblical literature. He had written a book, a copy of which was to be seen in his house, in which he had demonstrated, to his own satisfaction at least, that the institution of slavery was of divine origin. It was said that he was a brother of the Stringfellow who became so notorious during the Kansas troubles as a leader of the "border ruffians," who tried to force slavery into that territory before the breaking out of hostilities between the states.

Living at home with this Virginia doctor of divinity was a married daughter, whose husband was an officer in the Confederate army. They were people of the old school, cultured, refined and hospitable, though hard put to it to show any substantial evidences of their innate hospitality on account of their impoverished condition, which they seemed to feel keenly but were too proud to mention except when driven to it by sheer necessity. The Federal cavalrymen were always welcome in that house, and the officers in many instances were very kind to them. Indeed, I suspect that more than once they were spared the pangs of hunger by the thoughtful kindness of officers who had found shelter in their home and had broken bread at their table, only to suspect that the family larder had been stripped of the last morsel in order to keep up the reputation for Virginia hospitality.

About five miles farther down the river, in a lonely spot where a small tributary of the Rapidan tumbled down a decline, was a water power on which was a rude sawmill, where a single old-fashioned sash saw chewed its way lazily through hardwood logs. The mill was tended by its owner, who with his wife lived in a house hard by the mill, the only occupants of the dwelling and the only inhabitants of the immediate neighborhood. They led a lonely life, and when its monotony was broken by the arrival of the officer of the day upon his tour of duty, extended a quiet but what appeared to be a not over cordial welcome.

The man was a dwarf. He was so low in stature that when he stood, his head came just above the top of the dining room table. His diminutive stature was due to a strange malformation. His legs looked as if they had been driven up into his body, so that there was little left but the feet. Otherwise, he was like another, with well formed head and trunk. His wife was a comely lady both in form and in feature, rather above than below medium height. Both were intelligent and well read, pleasant people to visit with; but when this man, with the head and trunk of an adult, the stature of a child and to all intents and purposes, no legs at all, toddled across the floor; the effect was queer, and taken in connection with his somewhat solitary environment, it suggested a scene from the "*Black Dwarf.*" But when one was seated as a guest of these good people at their hospitable board, his physical deformity was lost sight of in the zest of his conversation.

The winter of 1863-64 was one of hard work for the Federal cavalry. In addition to their other duties, the Michigan regiments were required to change their tactical formation and learn a new drill. Up to that time, Philip St. George Cooke's single rank cavalry tactics had been used. The tactical unit was the set of fours and all movements were executed by wheeling these units. There was but one rank. For some reason, it was decided to substitute the old United States cavalry tactics and form in double ranks. The utility of the change was, to say the least, an open question, and it necessitated many weeks of hard and unremitting toil on the part of both officers and men. There was little time for rest or recreation. Long and tiresome drills and schools of instruction made up the daily routine.

In one respect however, these drills of troop, regiment and brigade were a good thing. Many hundreds of new recruits were sent on from Michigan, and being put in with the old men, they were worked into good soldiers before the campaign opened, and proved to be as reliable and efficient as the veterans with whom they were associated. The Sixth Michigan received over two hundred of these recruits at one time. They were fine soldiers, and on the march from the Wilderness to the James, no inspecting officer could have picked out the recruits of 1863-64 from those who enlisted in 1862.

At division and brigade headquarters alone was there time for play. Generals Custer and Kilpatrick had a race course where they used to devote some time to the sport of horse racing. There were in the division a number of blooded and speedy animals, and not a little friendly rivalry was developed in the various commands when the merits of their respective favorites were to be tested on the turf.

George Armstrong and Elizabeth Custer

It was while at Stevensburg that General Custer obtained leave of absence and went home to Michigan to claim his bride. He was married in February, 1864, to Miss Elizabeth B. Bacon, daughter of

Judge Bacon of Monroe, Michigan. Mrs. Custer accompanied him when he came back, and from that time on until the end of the war, whenever the exigencies of the service would permit, she was by his side. He was then but two months past twenty-four years of age, though he had already achieved fame as a cavalry officer and general of brigade. He was the youngest officer of his rank who won any great measure of success. Kilpatrick was more than three years his senior, although both were graduated from West Point in 1861.

Sometime after the beginning of the year 1864, there began to be rumors of some daring expedition that was on foot, to be led by the dashing general commanding the division. It was about the middle of February, when a number of statesmen of national prominence came to Stevensburg, and it did not take a prophet to tell that something of unusual importance was in the wind, though nothing very definite leaked out as to what it was. Among the visitors referred to were Senators Chandler (Zach) of Michigan, and Wilkinson of Minnesota.

During their stay, there was a meeting in a public hall in Culpeper at which speeches were made by both these gentlemen, and where General Kilpatrick demonstrated that he was no less an orator than a fighter. His speech was the gem of the evening and stirred up no end of enthusiasm. Hints were thrown out of an indefinite something that was going to happen. It is now known, as it was soon thereafter, that Kilpatrick had devised a daring scheme for the capture of Richmond, which had been received with so much favor by the authorities in Washington that he was then awaiting only the necessary authority from the War Department before setting out on what proved to be an ill-fated expedition.

Late in the month, permission was given and he proceeded to organize a force of picked men and horses, selected with great care from the various regiments. The Fifth, Sixth and Seventh Michigan and First Vermont were represented, the Sixth furnishing about three hundred men. The First Michigan had just re-enlisted at the expiration of its three years' term of service and was absent on veteran furlough, so did not take part, as the officers and men of that fine regiment would have been only too glad to do had they been given the opportunity. It was a small division, divided into two brigades. General Davies led one of them, but General Custer was taken away and entrusted with the command of an important diversion designed to attract the attention of the enemy by an attack on his left flank, while Kilpatrick passed around his right and by a quick march reached the Confederate capital. That portion of Custer's brigade which went on the raid, as it was called, was commanded by Colonel Sawyer of the First Vermont Cavalry. Detachments from the Fifth, Sixth and Seventh Michigan were commanded by Captain Hastings, Major Kidd and Lieutenant Colonel Litchfield respectively; the First Vermont by Lieutenant Colonel Preston.

Custer's part of the work was successfully accomplished. He created so much commotion in the direction of Charlottesville that Kilpatrick was across the Rapidan and well on his way before his purpose was either discovered or suspected. It was however, a fatal mistake to leave Custer behind. There were others who could have made the feint which he so brilliantly executed, but in a movement requiring perfect poise, the rarest judgment and the most undoubted courage, Kilpatrick could illy spare his gifted and daring subordinate; and it is no disparagement to the officer who took his place to say that the Michigan Brigade without Custer, at that time, was like the play of Hamlet with the melancholy Dane left out. With him, the expedition as devised might well have been successful; without him it was foredoomed to failure.

At the Culpeper meeting, there was a large gathering of both officers and enlisted men, attracted thither from various arms of the service by a natural curiosity to hear what the speakers had to say. There were also several ladies in the audience. On the platform sat many officers of high rank. I do not remember who presided, but recall distinctly the glitter of rich uniforms.

Culpepper, Virginia. Culpeper's strategic location made it a highly contested position for both the North and South. There were more than 160 skirmishes in and around Culpeper during the Civil War.

After the speaking had begun, an officer wearing the overcoat of an enlisted man came in from the wings and modestly took a seat at the back of the stage. "Not obvious, not obtrusive, but retired," he seemed to shun observation. When later he removed his overcoat, it was seen that he wore the dress uniform of a brigadier general. Inquiry disclosed that he was Wesley Merritt, commander of the Reserve Brigade of the First Cavalry Division. His brigade consisted of three regiments of regulars - the First, Second and Fifth United States Cavalry - and two regiments of volunteers - the First New York Dragoons and the Sixth Pennsylvania Cavalry. This was a crack brigade, and after the opening of the spring campaign it was closely associated with the Michigan Brigade for the remaining period of the war.

Wesley Merritt, whom I saw then for the first time, was one of the youngsters who received their stars in June, 1863. He was graduated from the West Point Military Academy in 1860 at the age of twenty-four, and made such rapid progress in rank and reputation that he was a brigadier at twenty-seven. As a cavalry commander, he was trained by John Buford. The latter was rightly called "Old Reliable," not because of his age, but for the reason that he rarely if ever failed to be in the right place at the right moment - solid rather than showy, not spectacular but sure. His courage and ability were both conspicuous. He belonged to the school of officers of which Thomas, Meade, Sedgwick and Gregg were exemplars, rather than to that of which Kearney, Sheridan and Custer were preeminent types.

Wesley Merritt. As colonel of the 5th Cavalry, Merritt was a member of the Court of Inquiry which first sat on January 13, 1879 presided over by Colonel John H King 9th Infantry to consider the alleged cowardice of Major Marcus A. Reno of the 7th Cavalry at the Battle of the Little Bighorn (June 25 to 26, 1876); which resulted in the death of General George Armstrong Custer.

Such also was Merritt, an apt pupil of an illustrious teacher, the lineal successor of Buford. He came by natural selection to be commander of the First Division, and at the last was Chief of Cavalry of the Army of the Potomac, the capable successor of

Pleasonton and Sheridan, a position for which he was peculiarly fitted by nature, by acquirements and by experience. Modesty which fitted him like a garment, charming manners, the demeanor of a gentleman, cool but fearless bearing in action, were his distinguishing characteristics. He was a most excellent officer, between whom and Custer there was, it seemed, a great deal of generous rivalry. But, in the association of the two in the same command there was strength, for each was in a sort the complement of the other. Unlike in temperament, in appearance and in their style of fighting, they were at one in the essentials that go to make a successful career.

But, to return to the point in the narrative from whence this digression strayed, the force that was thus assembled in Stevensburg, somewhat against the protests, but in compliance with orders from army and corps headquarters, was brought together with much show of secrecy, albeit the secret was an open one. As has been seen, the rumor of the projected movement had been for some time flying about from ear to ear and from camp to camp. Its flight however, must have been with heavy pinions, for it did not extend beyond the river where the Confederates were resting in fancied security, innocent of the hatching of a plot for sudden mischief to their capital.

The composition of the Second Brigade has already been given. Its numerical strength was about 1,800 officers and men. The First Brigade consisted of nine regiments of cavalry and one battery of artillery. That is to say there were detachments from that number of regiments. These were distributed equally among the three divisions as follows: From the First Division, the Third Indiana, Fourth New York and the Seventeenth Pennsylvania; from the Second Division, the First Maine, the Fourth Pennsylvania and Sixteenth Pennsylvania; from the First Brigade, Third Division, Davies's own command, the Second New York, the Fifth New York, and Eighteenth Pennsylvania. Ransom's regular battery was assigned to duty with this brigade.

The detachments from the First Division were all consolidated under Major Hall of the Sixth New York; those from the Second Division under Major Taylor of the First Maine. The aggregate strength of Davies's command was 1,817 officers and men, exclusive of the artillery. The total strength of Kilpatrick's command was about 3,500.

The expedition started after dark Sunday evening, February 28th, 1864, with three days' rations. The route selected led toward the lower fords of the Rapidan. The advance guard consisted of 600 picked men from the various commands, all under Colonel Ulric Dahlgren, an officer of Meade's staff who had established a reputation for extraordinary daring and dash. He had been especially designated from army headquarters to accompany the expedition. Davies followed with the main body of his brigade, including Ransom's battery. To Colonel Sawyer with the Vermont and Michigan men fell the irksome duty of bringing up the rear of the column, the chief care being to keep up the pace, not losing sight of those in front, of which for a good part of the night there was much danger.

The crossing was made a little before midnight at Ely's Ford, Dahlgren taking the Confederate picket post by surprise and capturing every man. No alarm was given. The start was thus auspicious. We were within the enemy's lines and they were not yet aware of it.

There was no halt. The rapid march was continued throughout the night. It was clear and cold. The order for the march was "at a fast walk," but every experienced cavalryman knows that the letter of such an order can be obeyed only by those in advance. The rear of the column kept closed up with great difficulty. The sound of hoofs in front was the only guide as to the direction to be taken. Often it was necessary to take the trot, sometimes the gallop, and even then the leaders were at times out of sight and out of hearing. At such times, there was an apprehensive feeling after the touch, which had to be kept in order to be sure that we were on the right road. This was especially true of the heads of subdivisions - the commanders of regiments - who were charged with the responsibility of keeping in sight of those next in front.

The march was not only rapid but it was continuous. There was an air of undue haste - a precipitancy and rush not all reassuring. Only the stoical were entirely free from disquietude. Those of us who were with the extreme rear, and who had not been admitted to the confidence of the projectors and leaders of the expedition, began to conjecture what it all meant, where we were going, and if the pace were kept up, when we would get there and what would be done when the destination was reached. All the excitement and enjoyment were Dahlgren's; all the dull monotony and nerve racking strain ours.

The head of column reached Spottsylvania Courthouse at daylight. The tail came trailing in as best it could sometime later. Here, in accordance with the prearranged plan, Dahlgren with his six hundred troopers separated from the main body, bearing to the westward and following the direct road to Frederickshall Station on the Virginia Central Railroad, his objective point being Goochland, about twenty miles above Richmond on the James River. The plan was for Colonel Dahlgren to cross the river at or near that place, move down on the south side, and be in position to recross by the main bridge into Richmond at ten o'clock Tuesday morning, March 1st, at the same moment when Kilpatrick would enter the city from the north by way of the Brook turnpike.[22]

But, "the best laid schemes o' mice and men gang aft a-gley." General Sheridan pointed out that such combinations rarely work out as expected, and that when an engagement with the enemy is liable to take place at any moment, it is better to keep the whole force well together.[23]

In this case for Kilpatrick, to divide his force was a fatal error of judgment. In the light of what took place it is now clear, as it ought to have been at the time that the entire command should have been kept together on one road. General Custer made the same mistake when he went to his death at Little Big Horn in 1876. The combination did not work out as he expected. It may be entirely safe and proper for detachments to be sent out to make diversions for the purpose of deceiving the enemy. This was done when on approaching Ashland Station, Major Hall was dispatched with a force of about five hundred men to drive in the pickets in front of that place and make a feint of attacking, leading the enemy to suppose that this was the main body, while Kilpatrick with most of his force proceeded without opposition on the road leading to Richmond. But care was taken that he could reunite at any moment.

It would have been better had Dahlgren continued as the advance guard, going directly to Richmond by way of one of the bridges of the South Anna River and the Brook, the main column closely following. In that way, the general commanding might have had all the parts of his expeditionary force well in hand, under his own eye, and there need have been no halting, hesitation or waiting one for the other. Dahlgren utterly failed to carry out to fulfillment the part of the plan prearranged for him to accomplish, and lost his life into the bargain.

And the pity of it is that his life was wasted. Had he died leading a charge through the streets of Richmond, compensation might have been found in the glory of his achievement. But he died in an ambush, laid for him by a small force of home guards and furloughed Confederate soldiers who managed to throw themselves across his way, when after admitted defeat, he was trying to make his escape with only a small portion of his command. He deserved a better fate.

The main body crossed the Po River in the morning of Monday, February 29th, and made a halt of fifteen minutes to feed. Thence it pushed on, Davies's brigade still leading by way of Newmarket, Chilesburg and Anderson's bridge across the South

Anna River to Beaverdam Station on the Virginia Central Railroad. This point was reached late in the afternoon, the rear guard not arriving until after dark. Here, some buildings and stores were burned.

A train coming into the station, warned by the reflection of the flames in the cloudy sky, backed out and escaped capture. A small force of Confederates made its appearance but was easily brushed away. The brushing and burning, however, were done by Davies's men. The Michigan cavalrymen coming too late for the fair were privileged to hover in the background and watch the interesting performance from a safe distance, leaving it for the imagination to picture what they would have done if they had had the chance.

This night was cold, raw and rainy, the atmosphere full of moisture which gradually turned to an icy sleet. This added greatly to the discomfort of the march, which was resumed after tearing up the track and taking down the telegraph wires and poles in the neighborhood of the station. The stop at Beaverdam Station was not worth mentioning so far as it gave any opportunity to men or horses for rest or refreshment. Out into the dark night - and it was a darkness that could be felt – rode those brave troopers. On and on, for hours and hours, facing the biting storm, feeling the pelting rain, staring with straining eyes into the black night, striving to see when nothing was visible to the keenest vision, listening with pricked up ears for the sound of the well shod hoofs which with rhythmical tread signaled the way.

The night was well advanced when at last a halt was ordered to make coffee for the men and give the patient animals the modicum of oats that had been brought, strapped to the cantles of the saddles. The bivouac was in the neighborhood of the Ground Squirrel Bridge. Davies in his official report said that he went into camp at eight o'clock in the evening. That may have been. Davies was at the head of column, and after the small advance guard, the first to reach the camp ground. It was fully two hours later when the last of the Second Brigade reached the place.

From seven o'clock Sunday evening until ten o'clock Monday night, there had been no stop to speak of - no chance to cook coffee or feed the horses - save the brief halt of barely fifteen minutes on the south bank of the Po River. The men were weary, wet, cold and hungry, but there was no complaining, for they were all hardened veterans, accustomed to hardship and exposure. They had been schooled to endure the privations of campaigning with cheerful fortitude.

When, at one o'clock Tuesday morning, March 1st, the march was once more resumed, it was found that the First Brigade still had the lead. As on the previous day, Michigan and Vermont were relegated to the rear. By the custom of the service, it was our turn to be in the advance. The rule was for brigades and even regiments to alternate in leading. That is because it is much easier to march in front than in rear. On that morning, Sawyer's command was entitled to be in front and the first in the fray. That may however, be looked upon as a trifling matter and not worth mentioning. Veterans will not so consider it. It was but natural that Kilpatrick should before all others have confidence in his old brigade and those officers with whom he had personally served. Davies was a gallant officer and had some fine officers and regiments with him. There were none better. It was an inglorious part that was assigned to us. Still, there was as it turned out not much glory in the expedition for anybody, least of all for Kilpatrick himself.

The march during the forenoon was along the Richmond and Potomac railroad, to and across the Chickahominy River, to the Brook turnpike. Davies advanced along the turnpike toward the city, driving in the pickets and capturing a few of them. He crossed the "Brook"[24] and succeeded in getting inside the outer entrenchments, within a mile of Richmond. From the high ground overlooking the intervening plain, it was almost possible to look into the streets and count the spires on the churches.

Richmond, Virginia. On April 4 and 5, 1865, President Lincoln made a conciliatory visit to Richmond as he pressed to conclude the war stating "with malice toward none, with charity for all." He was assassinated days later.

The time which it would take to make the ride from the Rapidan to the "Brook" had been closely calculated. Ten o'clock, Tuesday morning, March 1st, had been the hour set when Kilpatrick would arrive and begin the assault upon Richmond from the north, while Dahlgren attacked it from the south. The former was on time to the minute. But where was Dahlgren? He made no sign. There was no way to determine whether he was or was not carrying out his part of the prearranged plan. Signals did not work. Kilpatrick was left to his own resources. A condition had developed in which prompt decision and action were imperatively demanded. There was no time for delay or careful deliberation. To do or not to do, that was the question. And there was but one man who could settle it. The rationale of the raid was a hurried ride, timely arrival, great daring, a surprise, a sudden charge without a moment's hesitation - success.

Whatever was done must be done quickly. It was not conceivable that Kilpatrick with three thousand men and six pieces of artillery - Kilpatrick the bold, the dashing cavalryman, the hero of Middleburg and Aldie - the conceiver of the expedition, who knew in advance all about the perils he must meet, the chances he must take - that he would permit uncertainty as to what Dahlgren with but five or six hundred men and no artillery was doing to influence his own immediate action. For all that he knew, Dahlgren was already in position, ready to strike, but awaiting the sound of battle from the north as the signal to begin.

And yet he hesitated. The object of the expedition, as has been shown, was to ride into Richmond and liberate the prisoners. It was a daring enterprise. A courage to execute commensurate to the ability to conceive was presupposed. So far, everything had gone by the clock. Officers and men alike knew what that forced march of thirty-six hours without pause meant, if it had any rational meaning. Each one had screwed his courage to the sticking point to follow wherever our gallant commander led, prepared to share with him success or failure, according to the event.

Indeed, there was safety in following rather than in falling back. We were far afield in an enemy's country. It was necessary to hang together to avoid hanging separately. The goal was in sight. By a bold and quick forward movement alone could it be reached. An order to move up into a line of squadron columns was momentarily expected. That a dash into the city, or at least an attempt would be made, nobody doubted. Anything short of that would be farcical, and the expedition that set out big with promise would be fated to return barren of results. The good beginning was worthy of a better ending than that.

Well, some of Davies's advance regiments were dismounted and the men sent forward deployed as carbineers on foot to feel of the fortifications and make a tentative attack on their defenders. Some of Ransom's guns were unlimbered and opened fire at long range. Reply was made by the enemy's cannoneers, for some of the earthworks facing us were manned with artillerists.

In the meantime, Sawyer's brigade held on the pike in column of fours, mounted, anxiously awaiting orders and developments, listening intently to the desultory firing of the carbineers and the occasional boom of the cannon in front. There was a growing feeling of uneasiness and incertitude which began to frame our minds for doubts and fears as to the outcome.

At length, a staff officer was seen riding slowly from the front towards the rear. The thought that ran along the column was, "Now the order is surely coming to move forward at a trot." Not so, however. He had been directed by General Kilpatrick to notify commanding officers that in case any of their men should be wounded, they would be obliged to make their own arrangements for the transportation and care of them, since there were no ambulances available.

Cheerful intelligence, surely, and well timed to put men and officers upon their fighting mettle! From that moment, the mental attitude of the bravest was one of apathetic indifference. Such an announcement was enough to dampen the ardor of men as brave as those who had been selected to make up the personnel of this expedition.

Finally, anxious to get some idea of what was going on and what the outlook, I rode forward to a place overlooking the battlefield. Away to the front, a thousand yards or more, was an open stretch of cleared fields across which was a light line of dismounted cavalry skirmishers, firing away at the defenders of the earthworks. This defensive force did not appear to be formidable in numbers; nor was it particularly effective in its fire upon our troops. Along the Union line rode Captain L.G. Estes, adjutant general of the division, his cape lined with red thrown back on one shoulder, making of him a conspicuous target. He was exposing himself in most audacious fashion, as was his wont. It looked like an act of pure bravado. It was not necessary for him to furnish evidence of his gallantry. His courage was proverbial among the cavalrymen of the Third Division. They had seen him recklessly expose his life on many battlefields.

This was as near as the expedition ever came to capturing Richmond. Kilpatrick, who at the start was bold and confident, at the last when quick resolution was indispensable, appeared to be overcome with a strange and fatal irresolution. Davies was recalled and the entire force was directed to take the road to Meadow Bridge. It was after dark when we were ordered into camp somewhere between Mechanicsville and Atlee's Station. When I received the order, I inquired if we were to picket our own camp, but was informed that details for that purpose had been made and it would not be necessary. This quieted my fears somewhat but not entirely.

Precautions were taken against possible surprise and to ensure speedy mounting and getting into position in the event of an emergency requiring it. The regiment went into bivouac in line, a little back in the shadow and away from the fires. Few camp fires were permitted. The saddle girths were loosened slightly but the saddles were not removed. Each trooper lay in front of his own horse, pulling the bridle rein over his horse's head and slipping his arm through it. In this way, they were to get such sleep as they could. In case of a sudden alarm, they were to stand to horse and be ready instantly to mount.

Thinking that in any case it could be got ready while the regiment was being mounted, I allowed my own horse to be unsaddled and hitched him by the halter to a sapling in front of my shelter tent which was quickly pitched; Barnhart, the acting adjutant, and an orderly pitching theirs by the side of it. Then, removing sword and belt but keeping on overcoat, boots and spurs, I crawled in with a poncho under me, using the saddle for a pillow.

It was a raw, rainy night, and snow was falling. The bad weather of the first night out was worse than repeated. It seemed more like Michigan than Virginia. It was very dark. I do not believe that any man living could make a map of the camps which the two brigades occupied that night - the exact locations or even the relative positions of the various commands. I doubt if the actual participants could point them out were they to visit the place. I know that at the time, I had not the slightest knowledge on the subject and could not have told which way to go to find any one of them or even brigade or division headquarters.

It looked like a case of "wisdom consists in taking care of yourself." We were on the north side of the Chickahominy, and with the bridges guarded, it would be difficult for the forces with which we had been contending during the day to get in on our night encampment. At least they could not well take us by surprise. But this made the position all the more vulnerable from the north. It was idle to suppose that Stuart's cavalry was doing nothing. It was as certain as anything could be that his enterprising horsemen were gathering on our track, urging their steeds to the death in an endeavor to stop the audacious career of the Federal commander.

During the early evening, it was known throughout the command that the general had not given up the hope of capturing the city and liberating the prisoners. A body of five hundred men led by Lieutenant Colonel Addison W. Preston of the First Vermont Cavalry was to start out from our camp by the Mechanicsville Road, charge in, release the prisoners and bring them out, Kilpatrick covering the movement with his entire command. The latter's official report says there were two bodies, one to be led by Preston, the other by Major Taylor of the First Maine Cavalry. The name of Preston was a guarantee that the dash, if made at all, would be bravely led. There was no more gallant officer in the whole cavalry corps.

The conditions were such as to make one wakeful and alert, if anything could. But the danger of yielding for an instant to the allurement of the drowsiness produced by the long ride without sleep was overpowering. In an instant after getting under cover of the shelter tent, I was emulating the seven sleepers. It is doubtful if the trump of Gabriel himself, had it sounded, could have awakened me. The assurance that we were protected by pickets, and the order to go into camp having been given unaccompanied by any warning to be alert and on the watch for danger, had lulled me into such an absolutely false sense of security that I was for the time dead to all the surroundings. There was firing among the pickets. I did not hear it. A cannon boomed. I did not hear it. A second piece of artillery added to the tumult. I did not hear it. Shells hurtled through the trees, over the camp, and the waves of sound did not disturb my ear.

At last, partial consciousness returned. There was a vague sense of something out of the usual order going on. Then I found that Barnhart and the orderly were pulling me out of the pup tent by the heels. That sufficed. I was instantly wide awake. Barnhart was ordered to get his horse and mount the regiment, the orderly to saddle my horse and his own. In a few moments, all hands were in the saddle.

The regiment was wheeled by fours and moved a short distance to the right, more in the shadow and out of range of the shells, and formed in line facing toward where the enemy was supposed to be, and held there awaiting orders. No orders to advance came, nor was any brigade line of battle formed. In a very short time, a staff officer came riding fast and directed me to move out by fours on the road in rear of the alignment and follow the command which he said had gone and was retreating. He did not say what road it was nor whither it led. He then rode away.

Wheeling into column, the regiment was moved out on the road, and greatly confused as to the points of the compass, and not hearing or seeing anything of the column, turned in the wrong direction. The same staff officer soon overtook the head of the regiment and set us right. We had to countermarch, and as a matter of fact were going towards the enemy instead of joining in the retreat. It was by mistake, however. We had gone probably an eighth of a mile before being stopped.

The march then led back within sight of the camp which had been vacated. As we passed that point, far away in the distance among the trees by the light of the abandoned fires could be seen men flitting like specters through the places where the camps had been. They were presumably the enemy and apparently bent on plunder rather than conquest. It was a good time to give them a Roland for their Oliver, but there did not seem to be a disposition to make a concerted attack, or in fact, any attack at all. Kilpatrick was in full retreat toward Old Church, abandoning his plan of a midnight attack on Richmond.

The force which made the attack on the camps was led by Wade Hampton, who as soon as he knew of the expedition, set out on the trail, picking up odds and ends of Confederate cavalry when and where he could. He marched that day from Hanover Courthouse and says he came in sight of the camp fires near Atlee's Station and to his right on the Telegraph or Brook road. He must have been deceived as to the direction, for it is not possible that any portion of the main body could have been in camp on either of those roads. The camp he attacked was that of the Seventh Michigan, which bore the brunt of it. This regiment lost a number of prisoners, including the commanding officer, Lieutenant Colonel Litchfield.

We must have marched at least a mile, perhaps more, when the column was overtaken. It was moving at a walk on the road leading to Old Church. Finding myself in rear with no rear guard, I detached three troops (A, E and G) and held them with sufficient interval to cover the retreat. When there was a halt, they were formed in line across the road, and facing to the rear with carbines loaded and at a "ready" to repel any attack, should one be made. Once when halted, the tread of horses could be heard approaching. "Halt! Who comes there?" was the challenge. "Major Wells and a portion of the First Vermont Cavalry," was the reply. He advanced and was recognized, and for the remainder of the night we jointly looked after the rear until a camping ground was found near Old Church about daylight the next morning.

An amusing thing happened after Barnhart and the orderly pulled me out of the tent. The orderly saddled my horse, and after buckling on sword and belt, I put my foot in stirrup and proceeded to mount. The saddle slipped off to the ground. In the excitement, he had neglected to fasten the girths. I put the saddle on again, and making all tight, mounted and gave the horse the spur, when to my dismay he proved to be still tied to the tree. It was necessary to dismount, untie and adjust the halter. By this time, it is needless to say I was getting rattled. But the precautions taken made it easy to get the regiment into shape and keep it well in hand.

The most regrettable thing about it all was that Sawyer did not rush his entire brigade to the support of the picket line. Had that been done, it is more than likely that Litchfield and his men might have been saved from capture, though I do not know how Hampton found them when he stole into their camp. If they were scattered about and asleep, it would have been impossible to rally them and get them into line for effective resistance. On the other hand, had Sawyer with his other regiments or Davies with his brigade, or both of them together made a concerted attack, Hampton might have been worsted. But there was no attempt to make a fight. Hampton's attack caused consternation, forced a precipitate retreat, and led to the final abandonment of the objects of the expedition.

In a previous chapter, I have sought to show that official reports are often meager, sometimes misleading. There has always been a good deal of mystery about this affair. There is mystery still, which careful reading of the official records does not dispel. Sawyer made no report; or if he did, it was not published. Few if any of the regimental commanders submitted reports. The Michigan Brigade suffered its usual fate in that regard. Kilpatrick's report as published says: "The command was moved out on the road to Old Church, and placed in position and after considerable hard fighting repulsed the enemy and forced him back on the road to Hanover Courthouse."

Davies, in his official report said: "The enemy during the evening skirmished slightly with my pickets and about 12 p.m., attacked the Second Brigade in force. My command at once mounted and formed, but the Second Brigade unassisted repulsed the attack and I moved to the vicinity of Old Church." Davies, it is seen, did not claim to have made any fight. He was ready and in position, but moved away to Old Church.

Wade Hampton, who led the attack, says: "From Hanover Courthouse, I marched to Hughes's Crossroads, as I thought that would be the most likely place for the enemy to cross. From that place, I could see their camp fires in the direction of Atlee's Station as well as to my right on the Telegraph or Brook road. I determined to strike at the party near Atlee's, and with that view moved down to the station where we met the pickets of the enemy.

I would not allow their fire to be returned, but quickly dismounted 100 men and supporting them with the cavalry, ordered Colonel Cheek (of the North Carolina Brigade) to move steadily on the camp while two guns were opened on them at very short range. Kilpatrick immediately moved his division away at a gallop, leaving one wagon with horses hitched to it and one caisson full of ammunition. The enemy was a brigade strong here, with two other brigades immediately in their rear."

From these extracts it will be seen how commanding officers, when they write their official reports of a night rencounter, are apt to draw on their imaginations for the facts. The stout fight put up by Kilpatrick, and the graphic account by Hampton of how he whipped three brigades with a handful of Confederates hastily assembled, are equally mythical.

Davies's report gives a very accurate description of the affair.

From this, we find that he picketed toward Richmond and the Meadow bridges, taking care of the flanks and rear. The slight skirmishing with his pickets of which he speaks must have been with small bodies that came out from Richmond or which followed him from his position of the day on the Brook pike. It had no relation to Hampton's attack, which was from the opposite direction and entirely distinct. To Sawyer it was left, it would appear, to look out for the front - that is, toward Ashland and Hanover Courthouse. Sawyer sent the Seventh Michigan out on picket, the outer line advanced as far as Atlee's Station.

When Hampton came in from Hughes's crossroads, he did not stop to skirmish with the videttes. He did not fire a shot, but followed the pickets into the camp and opened with carbines and two pieces of artillery at close range. No arrangements appear to have been made to support the Seventh properly in the event of such an attack, which might have been foreseen. Sawyer should have reinforced the Seventh with his entire brigade. And it was equally incumbent on Kilpatrick to support Sawyer with Davies's brigade if he needed support. Neither of these things was done.

Kilpatrick's artillery made no response to that of Hampton. The only order was to retreat. Hampton was not far away from the facts when he said that "Kilpatrick immediately moved his division off at a gallop." He did not move it at a gallop. He moved it at a walk. But he moved immediately. He did not stop to fight, and morning found him well on the way to the Pamunkey River. It was an unlucky event for poor Litchfield. He was held as a prisoner of war very nearly if not quite until the curtain had fallen on the final scene at Appomattox. I do not remember that he ever again had the privilege of commanding his regiment.

Kilpatrick's strategy was better than his tactics. His plan was bold in conception, but faulty in execution. It has been shown that he made a mistake in dividing his command; that he made another when he failed to order an immediate attack after his arrival before the city. His afterthought of sending Preston and Taylor, at midnight, in a snow storm, and on a night so dark that it would have been impossible to keep together, to be sure of the way, or to distinguish friend from foe, to do a thing which he hesitated to do in the daytime and with his entire force, would have been a more serious blunder than either. Of course, if Preston had started, it would have been with the determination to succeed or lose his life in the adventure. That was his reputation and his character as a soldier. But the services and lives of such men are too valuable to be wasted in futile attempts. It might have been glorious, but it would not have been war.

The Jamestown Exposition was held from April 26 to December 1, 1907 in Norfolk, Virginia. It celebrated Jamestown, the first permanent English settlement in the present United States. The 20 remaining exposition buildings were included on the National Register of Historic Places in 1975, as a National Historic District.

To conclude this rambling description: In October, 1907, while attending the Jamestown Exposition, I met Colonel St. George Tucker, president of the exposition company and a well known scion of one of the first families of Virginia. The conversation turned to certain incidents of the Civil War, among others some of those pertaining to the Kilpatrick raid. Colonel Tucker was at the time a boy ten years of age. Armed with a gun, he was at a window in the second story of his father's house ready to do his part in repelling the "vandals" should they invade the streets of the city. This circumstance sheds light on the real situation. With the schoolboys banded together to defend their homes, and every house garrisoned in that way, not to mention the regular soldiers and the men who were on duty, it is quite certain that Richmond would have been an uncomfortable place that night for Preston and his little band of heroes. A man's house is his citadel, and boys and women will fight to defend it.

From Old Church, the command moved Wednesday to Tunstall's Station, and thence by way of New Kent Courthouse and Williamsburg to Yorktown. At Yorktown, the various regiments took transports to Washington, and from Washington marched back to their old camps around Stevensburg, no event of importance marking the journey. They arrived on the Rapidan about the middle of the month, having been absent two weeks. The men stood the experience better than the horses. The animals were weakened and worn out, and the time remaining before the opening of active operations was hardly sufficient for their recuperation.

THE WILDERNESS CAMPAIGN

IN the spring of 1864, the cavalry of the Army of the Potomac was thoroughly reorganized. Pleasonton, who had been rather a staff officer of the general commanding the army than a real chief of cavalry, was retired and Sheridan took his place. Kilpatrick was sent to the west and James H. Wilson, an engineer officer, succeeded him in command of the Third Division. Buford's old division, the First, was placed under Torbert, an infantry officer whose qualifications as a commander of cavalry were not remarkable. There were several of his subordinates who were both more capable and more deserving, notably Custer, Merritt and Thomas C. Devin. John Buford, the heroic, one of the ablest of all the generals of division, had succumbed to the exposures of the previous campaign. His death befell in December, 1863, on the very day when he received his commission as major-general, a richly deserved reward for his splendid and patriotic services in the Gettysburg and other campaigns. His death created a void which it was hard to fill. Gregg was the only one of the three old and tried division commanders who remained with the corps.

Of the generals of brigade, Merritt and Devin remained with their old division. Davies was transferred from the Third to the Second, and Custer's Michigan Brigade became the First Brigade of the First Division, the general going with it.

Pleasonton, who was sent to Rosecrans, in Missouri, although perhaps not like his illustrious successor a cavalry chief of the first rank, had a brilliant record and in the campaign of 1863 had performed most meritorious and effective service and certainly deserves a high place in the list of Union leaders of that period. In all the campaigns of the year 1863, he acquitted himself with the highest credit and in many of the battles, notably at Chancellorsville, Middleburg and Brandy Station, he was an equal match for Stuart and his able lieutenants.

General Sheridan (center, seated). Others pictured are General Wesley Merritt, General David Gregg, General Jefferson Columbus Davis and General Alfred Torbert.

If, in the readjustment incident to the assumption by General Grant of the chief command, Pleasonton could have been permitted to serve loyally under Sheridan, who was his junior in rank, it would doubtless have been better for both of them. He would have been obliged, to be sure, to crucify his ambition and waive his rank, but his name might have been linked with those of Gregg, and Merritt, and Custer in the record of "Little Phil's" picturesque marches from the Wilderness to the James; from Harper's Ferry to Cedar Creek; and from Winchester to

Appomattox. He left the army in whose achievements he had borne so honorable a part, and no opportunities for distinction came to him afterwards. Others wore the laurels that might have been his.

Soon after his arrival, General Sheridan reviewed the cavalry corps on the open ground near Culpeper. There were ten thousand mounted men in line, and when they broke into column to pass in review before the assembled generals of the army, it was a magnificent spectacle. To this day, the writer's blood quickens in his veins and a flush of pardonable pride mantles his face whenever he recalls the circumstance of one of Custer's staff, coming to his quarters after the parade to convey with the general's compliments the pleasant information that General Sheridan had personally requested him, to compliment the officers and men of the regiment on its excellent appearance and soldierly bearing on the review.

Only a short time before, General Kilpatrick had sent a similar message after seeing the regiment at brigade drill. How cheering these messages were; and how full of encouragement to the full performance of duty in the trying times that were close at hand! Life is not too full of such words of cheer, even when we do our best. It is not so much admiration as appreciation that one craves from his fellow men, especially from those who are by circumstance placed over him. But envy and malice, and a mean, begrudging spirit often stand at the door to keep it out, when it would fain enter, bringing the sunshine with it. There was nothing narrow or mean about Sheridan. Conscious of his own greatness, he was too broad to begrudge recognition to others. When a subordinate deserved commendation and Sheridan knew it, he always gave it.

Although the movement of the Army of the Potomac, which initiated in Virginia the campaign of 1864 and resulted in the battle of the Wilderness, began on May 3, it was the morning of May 4th when the Wolverine troopers left their camp near Culpeper. The

Second and Third Divisions, as has been shown, had the honor of leading the advance and preceded the infantry, crossing at Ely's and Germanna fords respectively on the day before. The First Division bivouacked on the north side of the river during the night of May 4th. At three o'clock on the morning of May 5th, the march was resumed, and crossing at Ely's ford, it moved to Chancellorsville and was encamped that night at the "Furnaces," south of the Orange Plank Road, about midway between Wilderness Church and Todd's Tavern, in the rear of the left of the Union lines.

Chancellorsville. The Battle of Chancellorsville is known as Robert E. Lee's "perfect battle" because his risky decision to divide his army in the face of a much larger enemy force, resulting in a significant Confederate victory. The victory was tempered by heavy casualties and the mortal wounding of General Stonewall Jackson to friendly fire, a loss that Lee equated to "losing my right arm."

Early on the morning of May 6th, "boots and saddles" and "to horse" summoned the brigade to arms; and at two o'clock a.m., it was on the march by the Furnace Road toward the intersection of that highway with the Brock turnpike. Gregg was at Todd's Tavern, at the junction of the Catharpin and Brock roads. Custer was to be the connecting link between Gregg's division and Hancock's corps. Devin, with the Second Brigade, was ordered to report to Custer. Wilson had been out the previous day on the Orange Plank Road and pike, beyond Parker's Store, where he encountered Stuart's cavalry and was roughly handled.

While moving up in the darkness, we came upon the scattered troopers of the First Vermont Cavalry, which for some time before the redistribution had been attached to the Michigan Brigade, but was then in Chapman's brigade of Wilson's division. They were moving to the rear, and seemed much chagrined over their defeat and declared that they did not belong to the Third Division, but were the "Eighth Michigan." "Come along with us," said their old Michigan companions-in-arms. "Wish we could," they replied.

Arriving at his destination before daylight, Custer posted his troops so as to be ready to meet the expected attack. Two troops, one from the First Michigan the other from the Sixth, commanded by Captain George R. Maxwell and Captain Manning D. Birge respectively, were sent well out on the Brock Road to picket the front. The line of battle was formed in the woods facing a cleared space, beyond which dense timber served as a screen to prevent the enemy's approach from being discovered. The right was held by the First and Sixth Michigan, formed in two lines, regimental front, the Sixth in rear, the men standing "in place, rest" in front of their horses. It was prolonged to the left by the Fifth and Seventh Michigan and Devin's brigade, composed of the Fourth, Sixth and Ninth New York and Seventeenth Pennsylvania regiments of cavalry. Devin however, did not arrive on the ground until the battle was well under way. The right of the line was "in the air," so far as was at that time known, the infantry not being in sight.

The open field directly in front extended some 200 yards beyond our position to the right, and it was perhaps 500 yards across it to the woods. The timber in which we formed extended from the rear clear around the right and across the front. In other words, the patch of open ground was enclosed on three sides, at least, by dense woods. The alignment faced in a westerly direction, and was back in the timber far enough to be hidden from the approaching foe. To the right and as it turned out somewhat to the rear lay the Army of the Potomac, which had been battling with Lee all the previous day; and orders had been issued for the fighting to be resumed at five o'clock in the morning.

Thus we stood, prepared in a state of expectancy, awaiting the sounds that were to summon us to battle. The brigade band was posted near the left flank of the First Michigan. General Custer, alert and wary, with a portion of his staff and escort, was out inspecting the picket line. The horse artillery had not yet arrived. Every trooper was alert and ready for whatever might come. The field, of which mention has been made, was bisected by a ravine, nearly diagonally from left front to right rear, the ground sloping into it from front and rear. This ravine was to play a prominent part in the battle that ensued.

Suddenly, the signal came. A picket shot was heard, then another, and another. Thicker and faster the spattering tones were borne to our ears from the woods in front. Then, it was the "rebel yell;" at first faint, but swelling in volume as it approached. A brigade of cavalry, led by the intrepid Rosser, was charging full tilt toward our position. He did not stop to skirmish with the pickets, but charging headlong drove them pell-mell into the reserves, closely following, with intent to stampede the whole command. It was a bold and brilliant dash, but destined to fall short of complete success. Rosser had met his match.

When the Confederate charge was sounded, Custer was near his picket line, and scenting the first note of danger, turned his horse's head toward the point where he had hidden his Wolverines in ambush, and bursting into view from the woods beyond the field, we saw him riding furiously in our direction. When he neared the edge of the woods, circling to the front and curbing the course of his charger as he rode, he bade the band to play, and with saber arm extended, shouted to the command already in the saddle; "Forward, by divisions!"

As the band struck up the inspiriting strains of "Yankee Doodle," the First Michigan broke by subdivisions from the right, the Sixth following in line, regimental front and the two regiments charged with a yell through the thick underbrush out into the open ground just as the Confederate troopers emerged from the woods on the opposite side. Both commands kept on in full career, the First and Sixth inextricably intermingled until they reached the edge of the ravine, when they stopped, the Confederates surprised by the sudden appearance and audacity of the Michigan men and their gallant leader; Custer well content with checking Rosser's vicious advance.

Some of the foremost of either side kept on and crossed sabers in the middle of the ravine. Among these was Lieutenant Cortez P. Pendill of the Sixth Michigan, who was severely wounded among the very foremost. One squadron of the Confederates, possibly a small regiment charging in column of fours, went past our right flank and then, like the French army that marched up a hill and then marched down again, turned and charged back without attempting to turn their head of column towards the place where Custer was standing at bay with his Michiganders clustered thick about him. Pretty soon, the Confederates ran a battery into the field and opened on us with shell. Every attempt to break Custer's line however ended in failure, the Spencer carbines proving too much of an obstacle to be overcome.

Meanwhile, the Fifth and Seventh had been doing excellent service on the left, forging to the front and threatening the right of the Confederate position. But it was evident that our own right was vulnerable, and Custer ordered Major Kidd to take the Sixth, move it by the rear to the woods on the right, dismount to fight on foot, and to use his own words; "Flank that battery."

The regiment had become much scattered in the charge, but the rally was sounded, and as many men as could be quickly assembled on the colors were withdrawn from the field, and obeying the order with as much alacrity as possible, in a few moments they were in position and moving forward briskly through the thick woods. But, they had not proceeded far when a strong line of dismounted Confederates was encountered. Both commanders seem to have ordered a simultaneous movement with a similar purpose, viz; to flank the other and attack his rear.

The two forces met very nearly on the prolongation of the line held by the mounted men of the First, Fifth and Seventh Michigan east of the ravine. The Confederate line extended beyond the right of the Sixth as far as we could see, and it was at once evident that we were greatly outnumbered, and liable to have the right flank turned at any moment. The little force stood bravely up to their work, using the Spencers with deadly effect and checking the advance of the Confederates in their immediate front. Major Charles W. Deane, who was helping to direct the movement, had his horse shot under him. Seeing that the left of the Confederates were trying to pass around our right flank, the captain of the left troop was directed to hold on to his position and the right was "refused" to protect the rear. At the same time, an officer was dispatched to General Custer with an appeal for reinforcements.

The entire of the Second Brigade was now up and a battery which arrived on the field after the withdrawal of the Sixth had been placed in position and opened upon the enemy. The battle was still raging in the field, but General Custer sent the Fifth Michigan, Colonel Russell A. Alger commanding, and the Seventeenth Pennsylvania, Lieutenant Colonel J.Q. Anderson commanding, to the relief of the Sixth Michigan. The reinforcements came none too soon. The Confederates, confident in their superior numbers, were pressing hard and threatening to envelop us completely.

In a solid line of two ranks, with Spencer carbines full shotted, the two magnificent regiments deployed into line on our right. Then, moving forward by a left half wheel, turned the tables on the too exultant foe, and he was forced slowly but surely back. By virtue of his rank, Colonel Alger was in command of the line, and in response to his clear voiced order, "Steady men, forward," the three regiments, with a shout, swept on through the woods, driving everything before them.

At the same time, the mounted men of the First and Seventh charged the force in their front. The enemy, thereupon, gave way in disorder, was routed and fled, leaving his dead and wounded in our hands. His repulse was complete and crushing, and we saw no more of him that day. The Michigan men, with the aid of Devin's New York and Pennsylvania troopers, had won a signal victory momentous in its consequences, for it saved the Union left from a disaster much dreaded, the fear of which neutralized one-half of Hancock's corps during the entire day.

No one who witnessed it can ever forget the superb conduct of Colonel Alger and his men when they swung into line on the right of the Sixth Michigan and turned a threatened reverse into a magnificent victory. Among the wounded, besides Lieutenant Pendill, already mentioned, were Captain Benjamin F. Rockafellow of the Sixth Michigan, and Lieutenant Alvin N. Sabin of the Fifth Michigan. All of these officers were severely wounded and all behaved with the most conspicuous gallantry.

In the meantime, what was the infantry doing? After Rosser was driven from the field, it was found that there was a line of infantry not far to the right and rear. Indeed, the left of the infantry line overlapped the right of the cavalry. Attention was called to the fact, when after the fight, some of the cavalrymen began to straggle to the rear, and returning said that the Twenty-sixth Michigan Infantry was only a little way off, and a good many of the men went over for a brief handshake with friends therein.

The Twenty-sixth Michigan was in Barlow's division. They had been interested listeners to, if not actual witnesses of the cavalry fight. The contest between the dismounted men of Rosser's and Custer's commands had been almost, if not quite, in their front and occasional shots had come their way. Why did not

Barlow or indeed Gibbon's entire command move up at the time when the Sixth Michigan Cavalry was contending alone with a superior force directly in their front? The answer to that question is in the sealed book which contains the reason of Grant's failure in the Wilderness. Let us see!

Grant's orders to the corps commanders - Sedgwick, Warren and Hancock - were to attack Lee's army at five o'clock a.m., May 6th. Longstreet had not arrived, but was expected up in the morning, and prisoners said he would attack the Union left. Hancock was directed to look out for the left. Barlow's division was posted for that purpose. Hancock's corps was divided into two wings, the right wing under Birney consisting of the three divisions of Birney, Mott and Getty; the left wing of Gibbon's and Barlow's divisions under Gibbon. Barlow, as has been seen, was to look out for the left. "The left" was well looked after by Sheridan's cavalry, for aside from Custer's two brigades which were directly in contact with Barlow's left flank, Gregg's division was posted at Todd's Tavern, still farther to the left.

Todd's Tavern. In 1864, The Union and Confederate cavalries clashed at the crossroads here in the race for Spotsylvania Court House.

Sedgwick and Warren attacked Ewell at the hour, but were

unsuccessful. Hancock's assault upon Hill was completely successful, although Longstreet arrived in the nick of time to save Hill. But Hancock's attack was with his right wing under Birney, and Longstreet struck the left of Birney's command. Where were the two divisions of Gibbon, posted for the very purpose of looking out for Longstreet?

In General A.A. Humphrey's, "*Virginia Campaigns*," page 40, we read: "At seven a.m., General Hancock sent a staff officer to General Gibbon, informing him of the success of his right wing and directing him to attack the enemy's right with Barlow's division. This order was only partially obeyed. Had Barlow's division advanced as directed, he (General Hancock) felt confident that the enemy's force would have been defeated. The cause of his failure was probably owing to the expected approach of Longstreet on his (Barlow's) left." Again: "At 8:30 a.m., Hancock began an attack with Birney's wing and Gibbon's division of the left wing."

General Grant, in his memoirs, (pp. 196-197): "Hancock was ready to advance, but learning that Longstreet was threatening his left flank, sent a division of infantry commanded by General Barlow to cover the approaches by which Longstreet was expected."

General Sheridan, (memoirs, vol. I, pp. 362-363): "On the sixth, General Meade became alarmed about his left flank and sent a dispatch saying: 'Hancock has been heavily pressed and his left turned. You had better draw in your cavalry to protect the trains'." And again: "On the morning of the sixth, Custer's and Devin's brigades had been severely engaged before I received the above note. They had been most successful in repulsing the enemy's attacks, and I felt that the line could be held. But the dispatch from General Hancock was alarming, so I drew all the cavalry close in around Chancellorsville."

Grant's memoirs, once more: "The firing was hardly begun when Hancock was informed that the left wing was seriously threatened so as to fully occupy Barlow. The enemy's dismounted cavalry opened on him (sic.) with artillery and pressed forward his skirmish line. The rapid firing of Sheridan's attack helped to confirm the impression that this was a serious flank attack by the enemy. These repeated reports prevented Hancock from throwing his full strength into the attack along the plank road."

"The rapid firing of Sheridan's attack" is good. Sheridan is entitled to the credit of placing Custer where he was. But that is all. Sheridan was not on the ground to direct the attack in any way; nor was the division commander on the ground. It was Custer's attack and it was Custer's victory. The only dismounted cavalry that attacked Barlow was Rosser's cavalry, and Custer's cavalry was between Rosser and Barlow. The only artillery with which the dismounted cavalry opened on Barlow was Rosser's battery and Custer and his men were between Barlow and that battery.

Had Barlow taken the trouble to ascertain what was really going on in his front, an easy matter, he would have found that, so far from this dismounted cavalry endangering his flank, they had been driven off the field in headlong flight, leaving their dead and wounded. There was never a moment during the entire day (May 6, 1864,) when Barlow was in the slightest danger of being flanked. His failure to advance enabled Longstreet to swing across his front and attack Birney's left, thus neutralizing Hancock's victory over Hill. If Barlow and Gibbon had advanced as they were ordered to do, they would have struck Longstreet's flank, and probably crushed it.

All of which seems to demonstrate that in battle, as in the ordinary affairs of life, imaginary dangers often trouble us more than those which are real. The fear of being flanked was an ever present terror to the Army of the Potomac, and the apparition which appeared to McDowell at Manassas, to Pope at the Second Bull Run, to Hooker at Chancellorsville, flitted over the Wilderness also, and was the principal cause why that campaign was not successful.

And then again, General Meade placed too low an estimate

upon the value of cavalry as a factor in battle and failed utterly to appreciate the importance of the presence of Sheridan's troopers upon his left. Had Meade and Hancock known Sheridan then, as they knew him a year later when he intercepted the flight of the Army of Northern Virginia at Five Forks and Sailor's Creek, there would have been in their minds no nervous apprehension that Longstreet might reenact in the Wilderness the part played at Chancellorsville by Stonewall Jackson. As it was, Grant's strategy and Hancock's heroism were paralyzed by these false rumors about Longstreet's menacing the safety of the Potomac army by moving against its left and rear. If such a thing was seriously intended, it was met and thwarted by Custer and Gregg, who alone and unaided as at Gettysburg, successfully resisted every effort on the part of Stuart's cavalry to break through the Union lines.

The noise of the successful battle which the Union cavalry was waging, instead of reassuring the Federal commanders as it should have done, served only to increase the alarm which extended to General Hancock and to army headquarters as well. If a proper rating had been placed upon the services of the cavalry, all apprehension would have been quieted. Barlow and Gibbon would have moved promptly to the front as directed, and Hill and Ewell might have been crushed before Longstreet was in position to save them.

General Sheridan's report gives a very meager and inadequate account of the cavalry fight in the Wilderness. In his book, he dismisses it with a paragraph. Major McClellan, Stuart's adjutant general, in his *"Campaigns of Stuart's Cavalry,"* makes no mention of it at all, though he devotes much space to Rosser's victory over Wilson on the fifth. That is not strange, perhaps, in the case of the Confederate chronicler, who set out in his book to write eulogiums upon his own hero, and not upon Sheridan or Custer. He has a keen eye for Confederate victories, and if he has knowledge of any other, does not confess to it.

As for Sheridan, his corps was scattered over a wide area, its duty to guard the left flank and all the trains, and he was not present in person when Custer put an abrupt stop to Rosser's impetuous advance. It is now known that he was so hampered by interference from army headquarters that his plans miscarried, and the relations between himself and his immediate superior became so strained that the doughty little warrior declared that he would never give the cavalry corps another order. By General Grant's intervention however, these difficulties were so far reconciled that Sheridan was soon off on his memorable campaign which resulted in the bloody Battle of Yellow Tavern and the death of the foremost Confederate cavalier, General J.E.B. Stuart.

THE YELLOW TAVERN CAMPAIGN

THE sequel to the false alarm about Hancock's left flank being turned was that all the cavalry was drawn in to guard the trains and protect the rear of the army. Custer's brigade moved back to the furnaces where it remained during the night. The morning of the seventh, he was ordered to resume his position of the day before. Gregg's division was returned to Todd's Tavern. Before the arrival of Gregg's command, the First Michigan Cavalry had a spirited encounter with Fitzhugh Lee in which Captain Brevoort, in command of the mounted men, particularly distinguished himself. There was pretty sharp fighting during the entire day, mostly on foot, the nature of the ground practically precluding movements on horseback.

The engagement of the cavalry on the seventh of May is known in history as the Battle of Todd's Tavern. It was made necessary in order to retake the position surrendered by Meade's order of the sixth. Much blood was shed and many valuable lives were lost in retrieving the error. In the events of the two days may be found a good illustration of the rule that an officer (even a great soldier like Sheridan) must obey orders, right or wrong. Sheridan must have known that there was no need to withdraw his cavalry from the left of the army. On the contrary, he knew that by all means it ought to remain where it was. Yet he obeyed and had to fight an offensive battle to regain what he was thus forced to give away. The conditions of the two days were reversed.

On the morning of the sixth, Sheridan was in possession and Stuart was trying to drive him out. On the morning of the seventh, Stuart was in possession and Sheridan had to drive him out. The material difference was that Stuart failed, Sheridan succeeded. Sheridan outgeneraled Stuart in both offensive and defensive tactics. The names of the respective chiefs are given here, but on the sixth, the actual fighting of the Union forces was directed by Custer and Gregg, of the Confederates by Rosser and Fitzhugh Lee; on the seventh, by Gregg, Merritt and Custer for the Federal side, by Fitzhugh Lee on the part of the Confederates. Gregg and Custer stood together in the Wilderness as they had done at Gettysburg. At Todd's Tavern, Merritt, Davies and Devin were added to the combination. And it was one that Stuart, Fitzhugh Lee nor Hampton was ever able to match.

At night, the First and Second divisions were encamped in the open fields east of Todd's Tavern, and in front of the positions held by them during the previous two days. Mounted pickets and patrols guarded the front and it soon became apparent that a movement of both armies was in progress. From front and rear came significant sounds which the practiced ear had no difficulty in interpreting. Grant, breaking off successively from his right, was passing by the rear to the left, concentrating around Todd's Tavern for a forward movement in the morning towards Spottsylvania Courthouse. The principle involved was to maneuver Lee out of the Wilderness into more open country by threatening his communications.

Once again, his strategic plans were thwarted by the faulty manner in which the tactics of the movement were executed. Sheridan had planned to seize Spottsylvania with his cavalry and his orders were for all three divisions to move at daylight with that end in view. Wilson was to lead and be followed up and supported by Merritt and Gregg with the First and Second Divisions. We shall see how Wilson was successful in carrying out his part of the plan, but how the others were stopped by orders from Meade, thus preventing the accomplishment of a well conceived enterprise and neutralizing two-thirds of the cavalry corps just when it was about to open the way to victory. By his peculiar tactical night movement, Grant held his line of battle intact except as the various corps broke successively from right to rear to march to the left. Thus Hancock's corps, though on the extreme left, was the last corps to move.

Spottsylvania Court House. The Battle of Spotsylvania Courthouse, was fought May 8–21, 1864. The battle, which ended in stalemate, included a brutal struggle over a section of the Confederate defenses that became known as the "Bloody Angle." The site of the Bloody Angle and other areas of the battlefield are preserved as part of Fredericksburg and Spotsylvania County Battlefields Memorial National Military Park.

Lee, quick to divine the purpose of his adversary, moved his army by the right flank on a parallel line. All night long, the ears of the alert cavalrymen could catch the indistinct murmur of troops moving with their impediments, which coming from both front and rear, bespoke the grand tactics of both commanders and presaged a great battle on the morrow. The pop, pop, pop of the carbines along the line of videttes was well nigh continuous, showing the proximity of the enemy's prowling patrols and scouts, and the necessity of constant vigilance. So closely did the Confederates approach the outposts that there was unceasing fear of an attack and neither officers nor men were able to obtain much rest. To sleep was out of the question. The First Michigan was held in readiness to make a mounted charge, while the other regiments were under orders to deploy dismounted, in case the attack which was looked for should be made. The officers of the First could be heard encouraging and instructing their men, keeping them alert and prepared for battle.

From the time of the organization of the Michigan Brigade, the First regiment had been designated as distinctively a saber regiment, the Fifth and Sixth for fighting on foot, as they were armed with Spencer rifles, and the result was that with them, dismounting to fight when in contact with the enemy in the early part of their terms of service became a sort of second nature. The First had a year's experience with the cavalry before the others went out, and it was in a saber charge at the Second Bull Run battle that Brodhead its first colonel was killed. The First Vermont, like the First Michigan, was a saber regiment and went out in 1861. When this regiment was attached to the brigade, Custer had three saber regiments, and it fell to the lot of the Fifth and Sixth Michigan to be selected more often than the others, perhaps for dismounted duty.

It often happened however, that the entire brigade fought dismounted at the same time; and sometimes, though not often, all would charge together mounted. Owing to the nature of the country, most of the fighting in Grant's campaign from the Wilderness to the James was done on foot. In the Shenandoah Valley campaign in the latter part of the year 1864, the reverse was the case, and at the battles of Tom's Brook, Winchester and Cedar Creek, the troopers in the command for the most part kept to the saddle throughout the engagements.

When Custer wanted to put a single regiment into a mounted charge, he generally selected the First Michigan, because it was not only older and more experienced but had many officers who possessed both great personal daring and the rare ability to handle men in action, keeping them well together so as to support each other and accomplish results. This regiment was not excelled by any other in the army for that purpose. The Seventh was an understudy for the First. The Fifth and Sixth worked well together on the skirmish line or dismounted line of battle and had no superiors in this kind of work. That they were pretty reliable when called upon mounted also is shown by the conduct of the Sixth in the Wilderness and of the Fifth at Trevillian Station.

It is only necessary to mention the gallantry of the Seventh at Hanovertown and at Yellow Tavern to demonstrate that it was an apt pupil of the First. All the officers and all the men of the Fifth, Sixth and Seventh took off their hats and gracefully yielded the palm to the First. It is doubtful if there was another regiment in the Federal cavalry service which contained so many officers highly marked for their fearless intrepidity in action. The circumstance of their talking to their men before an expected engagement was characteristic. They were always ready to face the peril and lead their men.

Later in the evening, away to the left where the infantry was going into bivouac, a Union band began to play a patriotic air. This was the signal for loud and prolonged cheering. Then a Confederate band opposite responded with one of their southern tunes and the soldiers on that side cheered. Successively, from left to right and from right to left this was taken up, music and cheering alternating between Federals and Confederates, the sounds receding and growing fainter and fainter as the distance increased until they died away entirely. It was a most remarkable and impressive demonstration under the circumstances, and lingered long in the memory of those who heard it.

Though the fighting on the 5th, 6th and 7th had been for the most part favorable to the Union troopers, it was disjointed and therefore neither decisive nor as effective as it might have been.

Sheridan believed that the cavalry corps should operate as a compact organization, a distinct entity, an integral constituent of the army, the same as the other corps. He looked upon his relation to the general in command as being precisely the same as that of Hancock, Sedgwick or Warren, and insisted that orders to the cavalry should be given through the cavalry corps commander just as orders to the Second Corps were given through General Hancock. He could not bring himself to consent to be a mere staff officer dangling at the heels of General Meade, but conceived himself to be an actual commander, not in name only but in fact.

Proceeding on this theory, he issued orders to the various division commanders to move at daylight on the morning of May 8th, and cooperate with each other under his personal direction in a plan which he had devised to seize Spottsylvania Courthouse in advance of Lee's infantry. They were to advance on converging roads in such a manner as to arrive successively but to support each other and open a way for the infantry columns. Wilson crossed Corbin's Bridge, charged through the town, driving out some of Fitzhugh Lee's cavalrymen and pursuing them several miles beyond. Merritt and Gregg made a good start, and if they had been allowed to proceed would have had no difficulty in accomplishing what Sheridan desired to have them do.

But without notice to Sheridan, Meade countermanded the orders to those two officers directing them to halt at the bridges and not cross. The result was that Wilson was isolated, Merritt's cavalry became inextricably entangled with Warren's infantry so that neither one of them reached Spottsylvania as they were both expected to do. Gregg was neutralized, Wilson's safety jeopardized, Sheridan's combinations broken up without his knowledge, and the way was left open for Lee's infantry, so that Anderson with Longstreet's corps took advantage of the situation and drove Wilson out and took possession - thus paving the way for Lee to form a defensive line there instead of farther south, probably inside the defenses of Richmond.

Then it befell that a series of bloody battles had to be fought to regain what was thus foolishly surrendered; to regain what indeed might have been held with slight loss, if Sheridan had been let alone and permitted to have his way. If he had been given a free hand, and assuming that Warren, Burnside, Sedgwick and Hancock would have carried out their part of the program with the same zeal and skill displayed by Sheridan, it is certain that the Battle of Spottsylvania with its "bloody angle" would never have taken place. The affair was a fiasco, but for that, no blame can be attached to either Sheridan or Grant, unless the latter be considered blameworthy for not directing the movements in person instead of leaving the tactics of the battle to be worked out by Meade.

Once more, as in the Wilderness, the cavalry was drawn in. The entire corps was massed in rear of the infantry and rendered inert. Sheridan with his ten thousand troopers was held idle and inactive while Warren, Sedgwick and Burnside were given the task of defeating Lee's veteran army without Sheridan's help. All his plans were rendered nugatory. He became satisfied that his efforts were useless. About noon, he went to Meade's headquarters and they had an interview which is one of the famous historical episodes of the Civil War. He told Meade that, inasmuch as his plans were to be interfered with, his orders countermanded, thus destroying the efficiency and usefulness of the cavalry corps, he must decline to give it further orders and General Meade could take it and run it himself, as he evidently desired to do.

He kept his poise however, sufficiently to intimate that he would like an opportunity to take his corps and go out after Stuart, since he believed he could whip Stuart in a fair fight if he could have a chance. Meade reported this conversation to Grant who told Meade to let him go and try. Grant had confidence enough in Sheridan to believe that he would make his word good.

The outcome of this was that the entire corps was ordered that very afternoon to concentrate at Alrich's, on the plank road leading to Fredericksburg, and be prepared to start at daylight on an expedition around Lee's right flank into the enemy's country. It was to be a second edition, only on a much larger scale and under a very different commander, of the Kilpatrick raid, an account of which was given in a previous chapter. The route selected was very much the same. But unlike Kilpatrick and others who had led cavalry expeditions up to that time, and whose idea was to ride rapidly through the country and avoid the enemy as much as possible, never fighting unless forced into it unwillingly, Sheridan went out with the utmost deliberation, looking for trouble - seeking it - and desiring before every other thing to find Stuart and fight him on his native heath. The confidence which he manifested in himself and in the prowess of his command was of its own kind, and a distinct revelation to the Army of the Potomac, in which it had long been a settled article of belief that Stuart was invincible, and indeed, up to that time he had been well nigh so, as Sheridan points out in his memoirs.

In the meantime, the battle was raging around Spottsylvania. Lee's army was getting into position, his various corps concentrating and entrenching, and making every preparation for a new base and a stout resistance. Grant's plans had all miscarried, thus far. Still, he had taken up his bridges and resolved to fight it out on that line. It was already evident that there was to be no more retreating. The officers and men of the Army of the Potomac made up their minds that they had crossed the Rapidan and the Rappahannock for the last time and that Lee would never be permitted to make a permanent halt outside the entrenchments of Richmond.

When the long column was marching along the rear of the army, the sounds of the battle going on could be distinctly heard. Hundreds of wounded men were coming from the front, mostly so slightly injured that they were helping themselves off the field to a place of safety where they could receive needed treatment. It filled us with astonishment to see the number of them. The official records show that Grant lost more than ten thousand men in the series of battles around Spottsylvania.

General Grant and staff

It seemed wicked to take ten thousand men well mounted and equipped away from the army at such a time as that. Queer ideas Meade had. And queerer still that Grant should have yielded to him in a matter of such vital importance. And the men that Sheridan was taking away were the very same troops with whom he broke Early's flank at Winchester; and who stood like a stone wall in the way of Early's advance at Cedar Creek after two corps of infantry had been routed only a few months later. Just imagine for a moment what might have been the result if Sheridan had been permitted to make the same use of his cavalry in the Wilderness or at Spottsylvania which he made of it at Winchester and Cedar Creek.

We camped at Alrich's for the night. And it was Sunday night. It will be remembered that the Kilpatrick expedition left Stevensburg on Sunday night. Three days' rations were drawn and issued to the men. There was but one-half of one day's ration of grain for the horses. So it was settled that our animals would have to depend on the country for their forage. The force thus assembled consisted of three divisions - about ten thousand troopers - under Merritt, Gregg and Wilson - seven brigades commanded by Custer, Devin, Gibbs, Davies, Irvin Gregg, McIntosh and Chapman. These were all veteran officers, often tried and never found wanting. Of these brigade commanders, two, Custer and Davies, held the rank of brigadier general; Devin was colonel of the Sixth New York; Gibbs of the First New York dragoons; Gregg of the Sixteenth Pennsylvania; McIntosh of the Third Pennsylvania; Chapman of the Third Indiana.

There were six batteries of artillery, all regulars but one - the Sixth New York Independent - Captain J.W. Martin. Pennington was still with the Third Division, as was the First Vermont Cavalry also. The four Michigan regiments were commanded by Lieutenant Colonel Peter Stagg, Colonel Russell A. Alger, Major James H. Kidd and Major Henry W. Granger, respectively.

The movement began at an early hour. The start was made long before daylight. General Custer, who was to lead, ordered that the Sixth Michigan move out first, and thus it fell to my lot to be in the van at the outset of that historic expedition. A guide was furnished with directions that the route taken be by the plank road to Tabernacle Church, thence to the Telegraph road running from Fredericksburg to Richmond, then due south toward Thornburg. The long column wound its way slowly out of the wilderness on a single road, marching by fours, Merritt in front, Gregg in rear, Wilson in the centre - seven brigades and six batteries - beyond doubt the most superb force of mounted men that ever had been assembled under one leader on this continent, and a more formidable body of horse than had been seen in that war on either side, up to that time, or was ever seen afterwards. The column when stretched out like a huge snake was thirteen miles in length, so that when the last of Gregg's regiments turned south on the Telegraph road, the head of Custer's brigade must have been nearing Chilesburg.

Telegraph Road

The night was clear and quiet; the air was soft and refreshing. To the right, the two great armies were sleeping. There was no note of bugle, no boom of cannon, no crack of rifle to disturb the tranquility of the night. As the dawn approached, the baying of dogs in the distance gave notice that the echoes of the march would soon reach the ears of the enemy's outposts.

But the morning was far advanced, the head of column well on its way past the right flank of Lee's army when the first hostile patrols were encountered. At a crossroad leading to the right, a small force of cavalry made its appearance. It was put to flight by Captain Birge with troop A. At this point, troop E, Captain A.E. Tower, was sent to the front as advance guard, Sergeant M.E. Avery with eight men going ahead with orders to charge any enemy that appeared on the road, the troop to follow him closely and the regiment to support the troop. General Custer with his staff and escort rode close up to the rear of the regiment.

Behind him came the other Michigan regiments, Devin's and Gibbs's brigades, then Chapman, McIntosh, Irvin Gregg and Davies in succession. Davies was to look out for the rear. Thus the latter, who led the Kilpatrick expedition, found his position reversed on this. The responsibility was great, and he met it with his accustomed courage and ability. Davies was one of the few men who early in the war found his niche and stuck to it. He was an ideal general of brigade; and he kept his place as such without a check until the war closed.

To those of us who had been with Kilpatrick but a short two months before, the contrast presented by a mental comparison of Sheridan's manner of conducting a march with that of his predecessor was most marked and suggestive. This movement was at a slow walk, deliberate and by easy stages. So leisurely was it that it did not tax the endurance of men or horses. There was a steadiness about it that calmed the nerves, strengthened self-reliance and inspired confidence. It was a bold challenge for the Confederates to come out and fight a duel to the finish. That they would be compelled to take up the gage thus thrown down there was no shadow of doubt.

Angelo Tower. Angelo Tower was the first man recruited by his good friend James H. Kidd (author) of Ionia, Michigan to serve in Company E, 6th Michigan Cavalry in 1862. His health was always poor and he resigned his in August 1864. He lived in Ionia, Michigan after the war.

The advance guard was kept active in the pursuit of Confederate scouts and pickets, small bodies of whom were constantly appearing in front or hovering on the flanks. Before reaching the point where the road leading to Beaver Dam was to be taken, the guide, either by ignorance or design, misled Avery and his men and took them to the eastward. Avery, suspecting something wrong, put a halter around the guide's neck and started to swing him up to the limb of a tree. He immediately discovered his mistake and a trooper was sent with word to take the other road, who reached the intersection just as the head of column did, so there was not a moment's delay. Avery soon came in with a squad of prisoners who with the guide were turned over to the provost guard. After reaching Chilesburg, we were on the same road over which we marched with Kilpatrick and needed no guide. The Confederate prisoners looked with astonishment upon this big body of cavalry which had stolen into their territory like a thief in the night, unexpected and unannounced.

During the day, as long as I had the advance, Captain Craig Wadsworth of Sheridan's staff rode by my side to represent and report to his chief. No very important incident happened, but the weather was pleasant, the air was exhilarating, the companionship was congenial, and there was sufficient of excitement to make it interesting. Things were kept moving and it was very enjoyable, as service with the advance of a marching column always is.

Late in the afternoon, we passed Chilesburg and the country began to have a familiar look. It was not yet dark when we crossed the North Anna River at Anderson's Bridge and the First Division prepared to bivouac on the south side. Gregg and Wilson went into camp for the night north of the river.

After crossing the river, Custer was ordered to proceed with his brigade to Beaver Dam Station. Here the First Michigan was given the advance, Major Melvin Brewer with one battalion as advance guard. The Sixth followed the First. Otherwise, the order of march was the same as during the day. A mile or so before

277

reaching Beaver Dam, Brewer came upon several hundred Union prisoners who were being hurried under the escort of Confederate infantry to the station, where trains were waiting to convey them to Richmond. His appearance, of course, resulted in the release of the prisoners, those of their guards who did not succeed in escaping by running away in the woods being captured.

The engineers began to sound their locomotive whistles as a signal for the Confederate escort to hurry up with their prisoners, and Brewer followed by the First and Sixth dashed into the station before the presence of the Michiganders was suspected, taking them by surprise and capturing the two locomotives with their trains. In a few minutes, Custer with the entire brigade was on the ground and it was found that besides the trains, he had captured an immense quantity of commissary, medical and other stores belonging to Lee's supply departments and which included nearly all his medical supplies. Everything that could not be carried away was destroyed. While this destruction was going on, some Confederates made their appearance in the adjacent woods and opened fire, but they were driven away without much trouble. This must have been a very severe loss to the Confederates.

The brigade then marched away and rejoined the division, every trooper having his horse loaded to the limit with such supplies as he thought he could use. General Merritt in his official report refers to this destruction of property as a mistake and characterizes the action as "gaucherie." It is, however, quite certain that the only way to have saved the supplies for issue to the corps would have been to move the division to Beaver Dam that night, for Stuart was concentrating his force at that point and might have been able to reclaim a portion of them if they had not been destroyed. At all events, Custer was on the ground and Merritt was not. Custer's action must have been approved by his judgment.

Early on the morning of May 10th, the march was resumed by the Negrofoot Road toward Ground Squirrel Bridge across the

South Anna River. It was even more leisurely than on the day before. Flankers were thrown out in both directions. The long column of fours thus proceeded slowly by the road while to the right and to the left, about 500 yards out, were parallel columns of flankers, marching by file, thus assuring that should the enemy attack either flank, it was only necessary to wheel by fours in that direction to be in line of battle with a very strong line of skirmishers well out in front.

But Stuart did not attack. He seems on that morning to have begun to comprehend Sheridan's plan, which was no doubt then sufficiently puzzling, but as we can see now, very simple. In a word, a slow and steady march straight toward the Confederate capital, all the time in position to accept battle should Stuart offer it. If he should not, to hold to the unyielding tenor of his purpose and with exasperating persistence continue to invite it. Stuart had turned off toward the east and was making a forced march with Fitzhugh Lee's division, consisting of the brigades of Lomax and Wickham, Gordan's brigade still hanging on to the rear of Sheridan's column. Our column made the march of eighteen miles to Ground Squirrel Bridge without molestation, and camped there that night on the south side of the river.

Stuart, after a much longer march, went into camp at Hanover Junction. At one o'clock in the morning, May 11th, he moved out toward Yellow Tavern, arriving there at about ten o'clock in the forenoon, before Sheridan's advance, which was headed in the same direction, made its appearance. Stuart had thus by a long and hard march brought his command where it could interpose between the Union cavalry and Richmond. He seems, however, to have been halting between two opinions - whether to form squarely across Sheridan's front or to hold his position on the flank until near enough to Richmond to be within reach of reinforcements from the troops that were being hurried into the city from the south to aid in the defense.

He appears to have chosen the latter alternative, for he

formed his command in a line running north and south, facing west, Wickham on the right, Lomax on the left with batteries near both his right and left flanks. The left of his line crossed the Telegraph road in front of Yellow Tavern where was quite an elevated piece of ground on which across the road was a battery well stationed and well manned. His men, however, must have been pretty well exhausted by the long march.

Yellow Tavern, which gave its name to the battle that ensued, is a hamlet at the junction of the Telegraph and Old Mountain roads, about six miles north of Richmond, where the first named road coalesces and becomes the Brook Turnpike, as I understand it. The Old Mountain Road comes down from the northwest, the Telegraph Road from the east of north. Sheridan struck the former at Allen's Station on the Fredericksburg railroad and followed it to Yellow Tavern. The Reserve Brigade reached that place a little before noon and finding Stuart in possession immediately began skirmishing. Devin came up next and was put on the line to reinforce Gibbs.

When Custer's brigade came up, pretty sharp skirmish firing could be heard in front. Merritt was in charge and the battle was on. Stuart had dismounted his entire force and formed them in a very strong defensive position on a commanding ridge beyond the tavern. Merritt had dismounted a portion of Gibbs's and Devin's commands and was feeling of Stuart's position. Custer's regiments, as they successively arrived, were massed mounted in column of battalions on the right of the road in a field, thus clearing the road. The march that day had been an easy one, the rest the night before had been complete, and never were men and horses in better condition or spirits for battle than were Sheridan's troopers.

Then there was an anxious pause. Glancing back, I saw that we were at the rear of the division. Down the road about 100 yards, a column of cavalry was approaching very slowly. Something at the head of the column attracted my particular attention and in a moment I made out that it was a general's battle flag. But I did not recognize it as one that I had seen before. There were a good many staff officers and a pretty large escort.

As they came opposite the regiment, the officer at the head looked back and saw that the flag was hanging limp around the staff, there not being air enough stirring to make it float out. He noted this and said to the color bearer, "Shake out those colors so they can be seen." The voice was mild and agreeable. The color-bearer did as directed and the general looked our way with a keen glance that was characteristic and took in every detail. Then instantly I knew who he was. I saluted and said, "Men, General Sheridan," and they gave him a cheer.

That was the first time I had seen Sheridan, except as I "looked toward" him when passing in review. One may do a good deal of service, even be in many skirmishes and battles, without getting a good look at the corps commander, much less the Commander of the Army. There was nothing about Sheridan's appearance at first glance to mark him as the principal figure in the scene. Except for the fact that he rode in front, one might have mistaken one of the other officers for chief. But close inspection easily singled him out. He was well mounted and sat his horse like a real cavalryman.

Though short in stature, he did not appear so on horseback. His stirrups were high up, the shortness being of leg and not of trunk. He wore a peculiar style of hat not like that of any other officer. He was square of shoulder and there was plenty of room for the display of a major general's buttons on his broad chest. His face was strong, with a firm jaw, a keen eye and extraordinary firmness in every lineament. In his manner there was an alertness evinced rather in look than in movement. Nothing escaped his eye, which was brilliant and searching and at the same time emitted flashes of kindly good nature. When riding among or past his troopers, he had a way of casting quick, comprehensive glances to the right and left and in all directions. He overlooked nothing.

One had a feeling that he was under close and critical observation, that Sheridan had his eye on him, was mentally taking his measure and would remember and recognize him the next time. No introduction was needed.

It would be as difficult to describe the exact physical traits that marked Sheridan's personality as to make a list of the characteristic mental attributes that distinguished him from others. There were perhaps no special, single, salient points. At least none were abnormally developed. In making an estimate of the man, it was the ensemble of his qualities that had to be considered. He had to be taken all in all. So taken, he was Sheridan. He was not another, or like another. There was no soldier of the Civil War with whom he fairly can be compared with justice to either. As a tactician on the field of battle, he had no equal, with the possible exception of Stonewall Jackson. In this respect, he to my mind more nearly resembled John Churchill, the great duke of Marlborough, than any other historical character of modern times of whom I have any knowledge. If he had not the spark of genius, he came very near to having it. This is a personal judgment put down here, the writer trusts, with becoming modesty and with no desire to put himself forward as a military critic.

Sheridan was modest as he was brave, reticent of his plans, not inclined to exploit his own merits, and he did not wear his heart or his mind upon his sleeve. His inmost thoughts were his own. What impressed us at this first sight of him was his calm, unruffled demeanor, his freedom from excitement, his poise, his apparently absolute confidence in himself and his troops, his masterful command of the situation. He rode away toward the front as quietly as he had come from the rear, with no blare of bugles, no brandishing of swords, no shouting of orders, no galloping of horses. In his bearing was the assurance that he was going to accomplish what he had pledged himself to do. He had found Stuart and was leisurely going forward to see for himself, to make an analysis of his adversary's position and so far as necessary to give personal direction to the coming conflict. But he was in no hurry about it and there was in his face and manner no hint of doubt or inquietude. The outcome was to him a foregone conclusion. Such was our chief and such was the beginning of the battle from which dates his fame as a cavalry leader and independent commander of the first rank.

Merritt and Custer were already at the front. Experience taught us that sharp work was at hand. It was not long delayed. The order came from General Custer for the Fifth and Sixth to dismount to fight on foot. The First and Seventh were held in reserve mounted. Not having visited this battlefield since that day, I am unable to give a very accurate description of its topographical features and shall not attempt to do so. The published maps do not throw a very clear light upon the matter, neither do the official reports. I am in doubt as to whether the Telegraph road and Brook turnpike are synonymous terms after passing Yellow Tavern or whether the former lies east of the latter.

As I have shown, Stuart's line ran along the Telegraph road, the right north of Half Sink, the left on a hill near Yellow Tavern. My authority for this is McClellan. Lomax held the left and had two pieces of artillery posted "immediately in the road;" one piece behind them "on a hill on the left." This would make his line extend due north and south, and our approach to attack it must have been from the west. Devin in his report says Stuart was driven off the Brook pike to a position 500 yards east of it. Whether that was at the beginning or near the close of the engagement is not quite clear. If the former, then the line referred to by Major McClellan could not have been on the Brook turnpike. I shall have to deal in general terms therefore, and not be as specific and lucid as I would like to be in describing Custer's part in the battle.

Just where the Michigan regiments were posted at the time they were ordered into the fight, I cannot say. They came down toward Yellow Tavern on the Old Mountain Road and I have no recollection of crossing the pike. It seems to me that they must have been west of it. We were moved across the road from where stationed when Sheridan came up, and deployed in the woods, the Sixth on the right of the Fifth. The line advanced and presently reached a fence in front of which was a field. Beyond the field and to the left of it were woods. In the woods beyond the field were the dismounted Confederate cavalry. Skirmishing began immediately across the field, each line behind a fence.

After a little, Captain Bayles of Custer's staff came from the right with an order to move the Sixth by the left flank and take position on the left of the Fifth. Just as he was giving this order, a great shout arose to the left, and looking in that direction, we saw that the entire of the Fifth Cavalry was climbing the fence and starting for a charge across the field. The Sixth instantly caught the infection, and before I could say "aye, yes or no," both regiments were yelling and firing and advancing on the enemy in the opposite woods. "You can't stop them," said Bayles. I agreed, and in a moment had joined my brave men who were leading me instead of my leading them.

The wisdom and necessity of Custer's order was however immediately apparent. Some Confederates lurking in the woods to the left opened fire into the flank of the Fifth Michigan, which for the moment threatened serious consequences. The line halted and there was temporary confusion. Quicker than it takes to tell it, Custer had appeared in the field mounted. One of Alger's battalions changed front and charged into the woods on the left, and the two regiments advanced and drove the enemy clear through and out of the woods in front. Barring the temporary check, it was a most gallant and successful affair, for which Custer gave the two regiments full credit in his official report. The line was then reformed with the Sixth on the left of the Fifth. At that time, this was the extreme left of the First Division and of the line of battle as well, the Third Division not yet having become engaged.

It was then found that the force with which we had been fighting had retreated to their main line of battle along a high ridge or bluff. In front of this bluff was a thin skirt of timber and a fence. Here Fitzhugh Lee's sharpshooters were posted in a very strong position indeed. Between the ridge and the edge of the woods where our line was halted was a big field not less than four hundred yards across, sloping down from their position to ours. To attack the Confederate line in front it would be necessary to advance across that field and up that slope. It looked difficult. The Confederate artillery was stationed to the right front on the extreme left of their line. We were confronted by Lomax"'s brigade. Beyond the right of the Fifth Michigan, Custer had the First Michigan, Colonel Stagg; the Seventh, Major Granger; and First Vermont, Lieutenant Colonel Preston; all mounted. They were across a road which ran at right angles with the line of battle, and in the direction of Lomax's battery.

As soon as our line appeared in the open - indeed, before it left the woods, the Confederate artillery opened with shell and shrapnel; the carbineers and sharpshooters joined with zest in the fray and the man who thinks they did not succeed in making that part of the neighborhood around Yellow Tavern an uncomfortably hot place, was not there at the time. It was necessary to take advantage of every chance for shelter. Every Wolverine who exposed himself was made a target of. Many men were hit by bullets. The artillerists did not time their fuses right and most of the damage was done to the trees behind us, or they were on too high ground to get the range. The line gradually advanced, creeping forward little by little until it reached a partial shelter afforded by the contour of the ground where it sloped sharply into a sort of ditch that was cut through the field parallel with the line of battle.

Here it halted, and the battle went on in this manner for a long time, possibly for hours. In the meantime, Chapman's brigade of Wilson's division had come into position on the left of the Sixth Michigan, thus prolonging the line and protecting our flank which until then had been in the air and much exposed. Off to the left, in front of Chapman, the lay of the land was more favorable. There were woods, the ground was more nearly level. The Confederate position was not so difficult of approach and gradually his left began to swing forward and threaten the right flank of Lomax's position, or more accurately, the Confederate center.

Thus, for several hours the lines faced each other without decisive results. At length, Sheridan determined upon an assault by mounted troops supported by those on foot. To Custer was assigned the important duty of leading this assault. It was toward four o'clock when Sergeant Avery, who had as quick an intuitive perception in battle as any man I ever knew, and whose judgment was always excellent and his suggestions of great value, called my attention to what appeared to be preparations for a mounted charge over to the right where General Custer was with his colors. "They are going to charge, major," said Avery, "and the instant they start will be the time for us to advance." That is what was done. The regiment forming for the charge was the First Michigan. Two squadrons under Major Howrigan led the vanguard. The bugles sounded "forward, trot, charge." Heaton's battery farther over was served with splendid effect.

Custer's staff passed the word along for the entire line to advance. There was no hesitation. The Fifth and Sixth and Chapman's regiments sprang forward with a shout. There was a gallant advance up the slope. Fitzhugh Lee's men held on grimly as long as they could, but there was no check to the charge. Howrigan kept on until he was among the guns sabering the cannoneers, capturing the two pieces in the road with their limbers and ammunition. In a few minutes, Custer and Chapman were in possession of the ridge and the entire line of the enemy was in full retreat. Back about 500 yards, the enemy attempted to make a stand and the Seventh Michigan was ordered to charge. This charge, led by Major Granger, resulted in his death. He was killed just before he reached the enemy's position, causing a temporary repulse of the regiment, but the entire line came on and the enemy was put to flight in all directions.

J.E.B. Stuart grave marker. Hollywood Cemetery, Richmond, VA.

Stuart was mortally wounded while trying in person with a few mounted men of the First Virginia Cavalry to stem the tide of defeat which set in when the First Michigan captured the battery. There is a controversy as to how he met his death. Colonel Alger claimed that Stuart was killed by a shot from one of the men on his dismounted line. Captain Dorsey of the First Virginia, who was riding with Stuart at the time, quoted by Major McClellan, says that he was killed by a pistol shot fired by one of the men who had been unhorsed in the charge on the battery, and who was running out on foot. In that case, it must have been a First Michigan[25] man, who very likely paid the penalty of his life for his temerity. It does not matter. One thing is certain. Stuart's death befell in front of Custer's Michigan Brigade, and it was a Michigan man who fired the fatal shot.

Stuart was taken to Richmond, where he died, leaving behind him a record in which those who wore the blue and those who wore the gray take equal pride. He was a typical American cavalryman - one of the very foremost of American cavaliers and it is a privilege for one of those who stood in the line in front of which he fell in his last fight to pay a sincere tribute to his memory as a soldier and a man. It fell to that other illustrious Virginian - Fitzhugh Lee - to gather up the fragments and make such resistance as he could to the further march of the Union cavalry.

YELLOW TAVERN TO CHESTERFIELD STATION

DAYLIGHT, May 12th, found the entire corps concentrated south of the Meadow bridges, on the broad tableland between Richmond and the Chickahominy River. Sheridan still kept his forces well together. Having accomplished the main purpose of the expedition - the defeat of Stuart - it remained for him to assure the safety of his command, to husband its strength, to maneuver it so as to be at all times ready for battle, offensive or defensive as the exigency might demand.

The next stage in the march of his ten thousand was Haxall's Landing on the James River, where supplies would be awaiting him. By all the tokens, he was in a tight place, from which all his great dexterity and daring were needed to escape with credit and without loss. His plan was to pass between the fortifications and the river to Fair Oaks, moving thence to his destination. Its futility was demonstrated when Wilson's division attempted to move across the Mechanicsville Road. It was found that all the ground was completely swept by the heavy guns of the defenses, while a strong force of infantry interposed. Reinforcements had been poured into Richmond, where the alarm was genuine, and it was clear that an attempt to enter the city or to obtain egress in the direction of Fair Oaks would bring on a bloody battle of doubtful issue.

Either course would at least invite discomfiture. To return by the Brook turnpike or Telegraph Road, even if that course could have been considered as an alternative, was alike impracticable. The cavalry force which had been trailing the command all the way from the North Anna River still maintained a menacing attitude in that direction. The only gateway out, either to advance or retreat, was by the Meadow Bridge over the Chickahominy, unless fords could be found. The river had to be crossed, and owing to the recent rains it was swollen.

Haxall's Landing. The buildings are Haxall's Four Mills on the James River.

All the signs pointed to a sortie in force from the fortifications. The defenders, emboldened by the hope, if not belief, that they had Sheridan in a trap; inspired by the feeling that they were fighting for their homes, their capital and their cause; and encouraged by the presence at the front of the president of the Confederacy - Jefferson Davis - were very bold and defiant, and even the lower officers and enlisted men knew that it was a question of hours at most when they would march out in warlike array and offer battle. Sheridan decided to await and accept it. Indeed, he was forced to it whether he would or not, as the sequel proved.

He sent for Custer and ordered him to take his brigade and open the way across the Chickahominy at the Meadow Bridges. Where work was to be done that had to be done, and done quickly and surely, Custer was apt to be called upon. The vital point of the entire affair was to make absolutely sure of that crossing, and Sheridan turned confidently to the "boy general" as he had done before and often would do again. The Michigan men were just beginning to stretch their limbs for a little rest - having fought all day the day before and ridden all night - when called upon to mount. They had not had time to prepare their breakfast or cook their coffee, but they rode cheerfully forward for the performance of the duty assigned to them, appreciating highly the honor of being chosen.

The road leading to Meadow Bridge descended to low ground and across the river bottoms. The wagon road and bridge were at the same level as the bottoms. Some distance below was the railroad. The grade for the track must have been at least twenty feet above the level where it reached the bridge which spanned the river. So the approach by the railroad was along the embankment.

When Custer reached the river, he found that the bridge was gone. The enemy had destroyed it. The railroad bridge alone remained. A force of dismounted cavalry and artillery had taken a position on the other side which commanded the crossing. Their position was not only strong but its natural strength had been increased by breastworks. Two pieces of artillery were posted on a slight hill less than half a mile back. In front of the hill were the breastworks; in front of the breastworks woods. A line of skirmishers firing from the edge of the woods kept the pioneers from proceeding with the work.

But Custer could not be balked. His orders were imperative. He was to make a crossing and secure a way for the entire corps to pass "at all hazards." He ordered the Fifth and Sixth Michigan to dismount, cross by the railroad bridge on foot and engage the enemy. The enemy's artillery swept the bridge, and as soon as it was seen that the Michigan men were climbing the railroad embankment to make the crossing, they trained their pieces upon it. Yet the two regiments succeeded. The Fifth led, the Sixth followed.

Damaged bridge over Chickahominy taken by famed Civil War photographer Matthew Brady.

One man, or at most two or three at a time, they tip-toed from tie to tie, watching the chance to make it in the intervals between the shells. Though these came perilously near to the bridge, none of them hit it, at least while we were crossing. They went over and struck in the river or woods below. It looked perilous, and it was not devoid of danger, but I do not remember that a single man was killed or wounded while crossing. It may have been a case of poor ammunition or poor marksmanship, or both. The worst of it was the nature of the ground was such that our artillerists could not bring their guns to bear.

Once over, the two regiments deployed as skirmishers and advancing with their 8-shotted Spencers, drove the Confederate skirmishers back through the woods and behind their breastworks, where we held them until a bridge was built, which must have been for two or three hours. The skirmishing in the woods was fierce at times, but the trees made good cover. It was here that Lieutenant Thomas A. Edie, troop A, Sixth, was killed by a bullet through the head. No attempt was made to assault the breastworks. The Confederates behind them, however, were kept so fully occupied that they were unable to pay any attention to the bridge builders, who were left unmolested to complete their work. This was the work which the two Michigan regiments were sent over to do, and they accomplished it successfully - something for which they never received full credit.

At one stage of this fight, my attention was attracted to the coolness of a trooper, troop A, Sixth, who was having sort of a duel with a Confederate. The latter was lying down in his works, the former behind a tree. When either one exposed any portion of his anatomy, the other would shoot. Some of the Confederate's bullets grazed the tree. The Michigan man would show his cap or something and when the other fired, step out, take deliberate aim and return the shot, then jump behind his natural fortress and repeat the maneuver. Finally, the Confederate ceased firing and there was little doubt that a Spencer bullet had found its mark. Making my way to the tree, I asked my man his name. His coolness and courage had much impressed me. "Charles Dean," he replied. "Report to me when the fight is over," I said. He did so, and from that day until the war ended, he was my personal orderly. A better, braver soldier or a more faithful friend, no man ever knew than Charles Dean, troop A, Sixth Michigan Cavalry.

After the completion of the bridge, the entire division crossed over. The Seventh Michigan, two regiments from Devin's brigade, two from Gibbs's - which with the Fifth and Sixth Michigan made seven in all, were put on the line as reinforcements, and an assault ordered. The entire line advanced, and even then it was no child's play. The Confederates fought well, but were finally driven out of their works and routed. Pursuit with dismounted men was useless. As soon as the horses could be brought over, the First Michigan and two of the Reserve Brigade regiments were sent in pursuit mounted, but were too late, most of the Confederates having made good their escape.

While this was going on, Gregg had a hard fight with the strong force of infantry and artillery which came out full of confidence to crush Sheridan. By a brilliant ruse, he took them by surprise and whipped them so thoroughly that they retreated within their inner fortifications, completely discomfited, and Sheridan remained on the ground most of the day with no one to molest or make him afraid.

Gregg's fight was characteristic of that fine officer who never failed to fill the full measure of what was required of him. Indeed, it was one of the most creditable actions of the war and one for which he never received full credit. The feeling throughout the First Division at the time, I know, was that the superb courage and steadiness of Gregg and his division had extricated Sheridan from a grave peril. The same Gregg, who with the help of Custer's Michigan Brigade, saved the Union right at Gettysburg, stood in the way and stopped a threatened disaster before Richmond.

After Gregg's repulse of the infantry, Custer's success in opening the way across Meadow Bridge and Merritt's rout of Fitzhugh Lee's cavalry, the Second and Third divisions remained unmolested for the rest of the day on the ground of the morning's operations, the First Division going to Gaines's Mills.

General Sheridan tells a story of two newsboys who came out after the fight with Richmond papers to sell. They did a thriving business, and when their papers were disposed of desired to return to the city. But they were so bright and intelligent that he suspected their visit involved other purposes than the mere selling of papers, and held them until the command was across the river, and then permitted them to go. There is an interesting coincidence between this story and the one told to the writer by St. George

Tucker of Richmond, and which appears on page 261 of this volume.

Late in the afternoon, the entire corps moved to Gaines's Mills and went into camp for the night. The march from Gaines' Mills to the James River was uneventful. When the head of the column on the 14th debouched on Malvern Hill, a gunboat in the river, mistaking us for Confederate cavalry, commenced firing with one of their big guns, and as the huge projectiles cut the air overhead, the men declared they were shooting "nail-kegs." The signal corps intervened and stopped this dangerous pastime.

Gaines' Mills. The Union defense at the Battle of Gaines' Mill produced seven Medals of Honor for soldiers who fought on June 27, 1862. Among those are Charles Hopkins, who would live until 1934, Ernest von Vegesack, a Swede who would later become a member of the Swedish Parliament, John Henry Moffitt, who would later become a Congressman from New York, and General Dan Butterfield, who would later compose the famous bugle call, "Taps."

Three days were taken here for rest, recuperation, drawing and issuing forage and rations, shoeing horses, caring for and sending away the sick and wounded, and in every way putting the command on a field footing again. It was a brief period of placid contentment. Satisfaction beamed from every countenance. Complacency dwelt in every mind. The soldiers smoked their pipes, cooked their meals, read the papers, wrote letters to their homes, sang their songs, and around the evening camp fires recalled incidents, humorous, thrilling or pathetic, of the march and battlefield. There was not a shadow on the scene.

On the 17th, the camp was broken and we marched by way of Charles City Courthouse, across the Chickahominy at Long Bridge to Baltimore Crossroads, arriving there on the evening of the 18th, when another halt was made. May 19th, I was sent with the Sixth Michigan to destroy Bottom's Bridge and the railroad trestle work near it. My recollection is that this was accomplished.

The next morning, General Custer was ordered with his brigade to Hanover Courthouse, the object being to destroy the railroad bridge across the South Anna River a few miles beyond. This necessitating a ride of more than twenty miles, an early start was made. The Sixth was given the advance, and it proved to be one of the most pleasant experiences of the campaign. The road led past Newcastle, Hanovertown and Price's; the day was clear, there was diversity of scenery and sufficient of incident to make it something worth remembering.

No enemy was encountered until we reached the courthouse. A small body of cavalry was there, prepared to contest the approach of the advance guard. The officer in command of the advance did not charge, but stopped to skirmish and the column halted. Foght, Custer's bugler, rode up and offered to show me a way into the station from which the Confederates could be taken in flank. Accepting his suggestion, I took the regiment and dashed through the fields to the left and captured the station, which brought us in on the left and rear of the force confronting the advance guard. Seeing this they took to flight, the advance guard pursuing them for some distance.

A quantity of commissary stores were captured here, some of which were issued to the men, the balance destroyed. The railroad track was torn up and two trestles destroyed where the railroad crossed the creek near the station. Custer moved his brigade back to Hanovertown and encamped for the night. The next morning, he returned to Hanover Courthouse, and sending the First and Fifth ahead, left the Sixth and Seventh to guard the rear. They advanced to near the South Anna River and found the bridge guarded by infantry, cavalry and artillery, which en route from Richmond to Lee's army had been stopped there for the exigency. Custer decided not to take the risk, as he learned that a force was also moving on his flank, and returned leisurely to Baltimore Crossroads.

Destroyed train near Richmond. Throughout the war, portions of the railroad were destroyed and rebuilt, and Confederates found it increasingly difficult to keep up with repairs. By the end of the war, the Confederate lines were almost completely unusable.

One incident of the first day seems to me worth narrating. The brigade bivouacked on a large plantation, where was a colonial house of generous proportions. It fronted on a spacious lawn, which sloped from the house to the highway and was fringed with handsome old spruce and Austrian pines. In front and rear the house had broad porches. A wide hall ran through the center of the house from one porch to the other and on either side of the hall were well furnished rooms of ample size. In rear, in an enclosure as broad as the house, was a well kept flower garden. It was a typical southern home of refinement and comfort. There were several ladies. The men were, of course, in the army.

General Custer with several of his officers called upon the ladies to pay his respects and assure them of protection. He was received with quiet dignity and refined courtesy, and for an hour chatted with them about the events then transpiring. They knew all the Confederate cavalry leaders and he was greatly interested in what they had to say about them. Before his departure, he left with one of the ladies a piquant and chivalric message for his friend Rosser, which she promised to deliver faithfully. Custer and Rosser, in war and in peace, were animated by the same knightly spirit. Their friendship antedated and outlived the war. The message was received and provoked one of a similar tenor in reply. He took especial care that no harm was done to the place, and marched away leaving it as good as he found it.

Upon our return, it was found that the Second and Reserve brigades, by the most extraordinary activity and skill, had succeeded in restoring the bridge across the Pamunkey at White House, on which the entire corps crossed over May 22nd. May 24th, Sheridan reported to General Meade at Chesterfield Station on the Richmond and Fredericksburg railroad, north of the North Anna River opposite Hanover Station. The two days' march from Aylett's was hot and dusty, and marked by nothing worth recalling, unless it be that the road after the cavalry had passed over it was dotted at regular intervals with the bodies of dead horses, the order having been that when horses gave out and had to be abandoned they must be shot.

HANOVERTOWN AND HAW'S SHOP

JUNE 26TH, the First and Second divisions, followed by Russell's division of the Sixth corps, started down the north bank of the Pamunkey River to secure the crossings, Grant having determined on another movement by the left flank, and to throw his entire army across into the territory between the Pamunkey and Chickahominy. Feints were made that day at the fords near Hanover Courthouse, but after dark both Torbert and Gregg, leaving a small force on duty at each of these fords respectively, quietly withdrew and made a night march to Dabney's Ferry opposite Hanovertown, the First Division leading. At daylight, Custer in advance reached the Ferry and the First Michigan under Colonel Stagg gallantly forced the passage, driving away about one hundred cavalrymen who were guarding it and making a number of them prisoners. The entire division then crossed and moved forward through the town.

General Custer directed me to take the road from Hanovertown and push on in advance toward Hanover Courthouse. We had gone but a mile or so when in the midst of a dense wood, a force which proved to be dismounted cavalry was encountered, strongly posted behind temporary earthworks hastily thrown up. The regiment was dismounted on the right of the road, the First Michigan, following closely, went in on the left and the two regiments made a vigorous attack, but met with a stubborn resistance and did not succeed in carrying the works at once. A band was playing in rear, indicating the presence of a brigade, at least.

Noticing that a portion of the enemy's fire came from the right, I sent the sergeant major to the rear with word that the line ought to be prolonged in that direction. The non-commissioned officer returned and reported that the message had been delivered to the brigade commander, but that it was overheard by the major general commanding the division, who exclaimed with a good deal of impatience; "Who in ---- is this who is talking about being flanked?" I was mortified at this, and resolved never again to admit to a superior officer that the idea of being flanked had any terrors. But General Torbert, notwithstanding, did reinforce the line with a part of General Devin's brigade in exact accordance with my suggestion.

Custer, however, did not wait for this, but taking the other two regiments of his brigade (the Fifth and Seventh Michigan) made a detour to the left by way of Haw's Shop, and came in on the flank and rear of the force which the First and Sixth, with Devin's help were trying to dislodge from its strong position, and which held on tenaciously so long as it was subjected to a front attack only. But, as soon as Custer made his appearance on the flank, the enemy, Gordon's brigade of North Carolinians, abandoned the earthworks and fled, the First and Sixth with Devin's regiments promptly joining in the pursuit.

Custer's approach was heralded by an amusing incident. The band that had been challenging us with its lips of brass stopped short in the midst of one of its most defiant strains, and the last note of the "*Bonnie Blue Flag*" had scarcely died on the air, when far to the left and front were heard the cheery strains of "*Yankee Doodle.*"[26] No other signal was needed to tell of the whereabouts of our Michigan comrades, and it was then that the whole line moved forward, only to see as it emerged into the open, the Tar-heels of the South making swift time towards Crump's Creek, closely followed by Custer and his Michiganders. The latter had accomplished without loss by the flanking process what he had tried in vain to do by the more direct method.

The charge of the Fifth and Seventh Michigan, commanded by Captain Magoffin and Major Walker respectively, and led by General Custer in person, was most brilliant and successful, the Seventh continuing the pursuit for about three miles. First Sergeant Mortimer Rappelye of troop C, Sixth, and one of his men were killed at the first fire. Rappelye was in command of the advance guard and had been slated for a commission, which he would have received had he lived.

That night, the cavalry encamped on Crump's Creek. The next day, the army was all over and Grant had taken up a new line extending from Crump's Creek to the Totopotomoy. Still, he was uncertain of what Lee was doing, and it became necessary to find out. This led to what was one of the most sanguinary and courageously contested cavalry engagements of the entire war - the Battle of Haw's Shop - in which Gregg and Custer with the Second Division and the Michigan Brigade, unassisted, defeated most signally, two divisions under the command of Wade Hampton in his own person. Indeed, it is not certain that it was not even a more notable victory than that over Stuart on the right flank at Gettysburg. It was won at a greater sacrifice of life than either Brandy Station or Yellow Tavern.

After the death of Stuart, though so short a time had elapsed, the Confederate cavalry had been reorganized into three divisions commanded by Wade Hampton, Fitzhugh Lee, and W.H.F. Lee, the first named being the ranking officer. His division had been largely reinforced, notably by a brigade of South Carolinians under M.C. Butler, who after the war was the colleague of Hampton in the United States Senate. This brigade consisted of seven large regiments, numbering in all about four thousand men. It was a brigade that honored the state which produced Sumter, Marion, the Rutledges and the Hamptons.

All this cavalry had joined the army of Northern Virginia and was in position to cover the movements which Lee was making to confront the Army of the Potomac. Sheridan's corps, now that it had returned to the army, was once more somewhat dispersed. Wilson was still north of the Pamunkey, covering the transfer of the several infantry corps and guarding the fords. The First Division, as we have seen, led the crossing on the 27th, and was covering the front and right of the infantry along Crump's Creek. Gregg, who had followed Torbert, was at Hanovertown.

On the morning of May 28th, Gregg was sent out by Sheridan to discover the movements of Lee, who was skillfully masking his designs behind his cavalry. Gregg had advanced but a short distance beyond Haw's Shop when, in a dense wood protected by swamps, behind breastworks of logs and rails and with batteries advantageously posted, he found the enemy's cavalry dismounted and disposed in order of battle. He promptly attacked, notwithstanding the disparity in numbers and in position, Davies going into action first, followed by Irvin Gregg, and the entire division was quickly engaged.

Gregg was resolute, Hampton determined, and for hours the battle was waged with the most unyielding bravery on both sides. The list of killed and wounded was unexampled in any other cavalry contest of the Civil War, aggregating in the Second Division alone two hundred and fifty-six officers and men. Davies's brigade lost twenty-three officers. The First New Jersey Cavalry had two officers killed and nine wounded. The enemy's losses were even greater.

It was an unequal contest - one division against two, two brigades against four - with the odds in favor of the Confederates. Hampton, who in the beginning maintained a posture of defense, began to assume a more aggressive attitude and showed a disposition to take the offensive. In the afternoon towards four o'clock, he brought up Butler's brigade to reinforce the center of his line. These troops were armed with long range rifles and many of them had not been under fire before. This was their first fight.

They came on the field with the firm purpose to win or die, and preferred death to defeat or surrender, as the sequel proved.

Then and not until then, it began to look as though the hitherto invincible Gregg might have the worst of it. There was danger that the center of his line would be compelled to yield. It was in front of this new and valorous foe that the First New Jersey suffered its fearful losses. The attack was such that only the bravest men could have withstood it.

At this critical juncture, Sheridan ordered Custer to the front to reinforce Gregg. It was time. The Michigan men were having a rest, thinking it was their turn for "a day off." But, as in the Wilderness and at Meadow Bridge, they were instantly in the saddle and en route. Marching by fours along a country road, hearing the sounds but not yet within sight of the conflict, lines of Federal infantry were seen marshaled for action, and a knot of officers of high rank gazing toward the front. Passing to the right of these, the column turned to the right into the road leading past Haw's Shop, and through the woods where the two lines were fiercely contending, and which road bisected the battlefield.

An impressive scene came into view. Beyond the wood, less than a mile away, which extended on both sides of the road, one of Hampton's batteries was firing shell with the utmost rapidity. These shells were exploding both in the woods and in a broad plain behind them and to the right of the column as it advanced. Hundreds of non-combatants were fleeing to the rear across this open space. The woods, like a screen, hid the battery from view. Only the screaming and exploding shells could be seen. When the head of the Michigan column came into their line of vision, the Confederate cannoneers trained one of their guns on the road and the shells began to explode in our faces. A right oblique movement took the column out of range.

Gregg's men had been gradually forced back to the very edge of the woods, and were hanging on to this last chance for cover with bull dog tenacity. The enemy were pressing them hard, and apparently conscious that reinforcements for them were coming, seemed to redouble their fire both of artillery and small arms. It was a fearful and awe inspiring spectacle.

Confederate General Wade Hampton. After the war, Hampton gave tacit support to the Ku Klux Klan, which had independent chapters arising throughout the South. He helped raise money for legal defense funds after the government started to enforce anti-Klan legislation to suppress the violence against freedmen and white Republicans.

Custer lost no time. Massing the brigade close behind Gregg's line of battle, he dismounted it to fight on foot. Every fourth man remained with the horses which were sent back out of danger. The line formed in two ranks like infantry. The Sixth was to the right, its left resting on the road; the Seventh to the left, its right on the road. The First formed on the right of the Sixth, the Fifth on the left of the Seventh. The time for action had come. It was necessary to do one thing or the other. No troops in the world could have been held there long without going forward or back.

Custer, accompanied by a single aide, rode along the line from left to right, encouraging the men by his example and his words. Passing the road, he dashed out in front of the Sixth, and taking his hat in his hand, waved it around his head and called for three cheers. The cheers were given and then the line rushed forward. Custer quickly changed to the flank, but though thus rashly exposing himself, with his usual luck, he escaped without a scratch. Christiancy, his aide, had his horse shot under him and received two wounds, one a severe one through the thigh.

Gregg's men permitted the Michigan men to pass. In a moment, the Wolverines and the Palmetto men were face to face and the lines very close. Michigan had Spencers. South Carolina, Enfields. Spencers were repeaters, Enfields were not. The din of the battle was deafening. It was heard distinctly back where the infantry was formed and where Grant, Meade and Sheridan anxiously were awaiting the event. The Spencers were used with deadly effect. The South Carolinians, the most stubborn foe Michigan ever had met in battle, refused to yield and filled the air with lead from the muzzles of their long range guns as fast as they could load and fire.

The sound of their bullets sweeping the undergrowth was like that of hot flames crackling through dry timber. The trees were riddled. Men began to fall. Miles Hutchinson, son of my father's foreman who had left home to go to the war with me, fell dead at my side. Jimmie Brown, the handsome and brave sergeant, dropped his piece, and falling, died instantly. Corporal Seth Carey met his fate like a soldier, his face to the foe. A member of troop H, shot through the breast, staggered toward me and exclaiming, "Oh, major," fell literally into my arms, leaving the stains of his blood upon my breast.

This strenuous work did not last long. It may have been ten minutes from start to finish - from the time we received the South Carolinians' fire until the worst of it was over and they began to give way. But, in that brief ten minutes, eighteen brave men in the ranks of the Sixth Michigan had been either killed or mortally wounded; and as many more were wounded but not fatally. The enemy suffered even more severely. The brigade lost forty-one

killed - eighteen in the Sixth; thirteen in the
Fifth; five in the First and five in the Seventh. The losses of the Fifth in officers and men wounded but not fatally were larger than those in the Sixth, the total of killed and wounded aggregating something like fifty in the regiment. The First, though it did not meet with so sturdy a resistance in its immediate front, was able to work around the flank of the enemy, thus materially aiding in breaking their spirit and putting them to rout.

Some of the South Carolina men exhibited a foolhardy courage never seen anywhere else so far as my knowledge extends. "Surrender," said Sergeant Avery to one of them who had just discharged his piece and was holding it still smoking in his hands. "I have no orders to surrender - you," returned the undaunted Confederate. He surrendered, not his person, but his life.

Such a fate befell more than one of those intrepid heroes. It was a pity, but it was war and "war is hell." The enemy's line at that time had been driven beyond the woods into a clearing where was a house. While crossing a shallow ravine before reaching the house, it was noticed that shots were coming from the rear. An officer with a troop was ordered back to investigate. It was found that at the first onset, the regiment had obliqued slightly to the right, thus leaving an interval between the left flank and the road in consequence of which about fifteen Confederates had been passed unnoticed. Some of them had the temerity to begin giving us a fire in the rear. They were all made prisoners.

The force in front was driven from the field, leaving their dead and wounded. Eighty-three dead Confederates were counted by those whose duty it was to bury the dead and care for the wounded in the field and woods through which the Michigan men charged. Those who were killed in front of the Sixth Michigan were South Carolinians from Charleston, and evidently of the best blood in that historic city and commonwealth. They were well dressed and their apparel, from outer garments to the white stockings on their feet, was clean and of fine texture. In their pockets, they had plenty of silver money.

In this engagement, as well as in that at Hanovertown the day before, the Fifth Michigan was commanded by Captain Magoffin, Colonel Alger having remained at White House for a few days on account of illness. Colonel Stagg and Major Alexander Walker led the First and Seventh respectively.

General Sheridan narrates that when he called upon Mr. Lincoln in Washington the president made a facetious reference to General Hooker's alleged fling at the cavalry, when he asked; "Who ever saw a dead cavalryman?" It is perhaps doubtful whether Hooker uttered so pointless a saying, devoid alike of sense and of wit. If such a question was ever seriously propounded by him or by anyone else, its sufficient answer could have been found upon the battlefield of Haw's Shop. And not there alone. The First Michigan cavalry had sixteen killed, including its colonel at the second Bull Run and twelve at Gettysburg. The Fifth Michigan lost fifteen killed at Gettysburg; the Sixth Michigan twenty-four at Falling Waters and the Seventh Michigan twenty-two at Gettysburg - all of these before General Sheridan had that interview with Mr. Lincoln in the White House.

This record was enough of itself to render the cavalry immune to ironical disparagement. If there were any honest doubts as to the efficiency and fighting qualities of the Potomac cavalry, they were dissipated by the campaign of 1864. After Todd's Tavern, Yellow Tavern, Haw's Shop, Cold Harbor and Trevilian Station, no slurring remarks aimed at the cavalry were heard. Its prestige was acknowledged in and out of the army by all those who had knowledge of its achievements and were willing to give credit where credit was deserved.

An all night march followed the battle, after the dead had been buried and the wounded cared for. The morning of May 29th found the two divisions in the neighborhood of Old Church and thence in the afternoon of May 30th, Custer and Merritt marched out toward Cold Harbor, the Reserve brigade in advance to reinforce Devin, who was having a hot fight at Matadequin Creek with Butler's South Carolinans posted on the opposite side in a strong position. The entire division became engaged, the fighting being mostly dismounted and the opposing force was driven in great confusion from the field. The Sixth Michigan was held in reserve mounted and expected to be ordered in for a mounted charge, but for some unexplained reason the order did not come. The First, Fifth and Seventh were in the thickest of it and rendered excellent service. The pursuit was kept up for several miles, and the enemy retreated to Cold Harbor, leaving his dead and wounded on the field as at Haw's Shop. Butler's men behaved with great gallantry, but were ready to surrender when the logic of the situation demanded it. They made no such resistance as in the former action.

May 31st, in the afternoon, the First Division advanced on Cold Harbor, Merritt in advance, on the road leading from Old Church. Custer followed Merritt. Devin was sent by another road to the left with the intention of having him attack in flank the force which the other two brigades were engaging in front. The Sixth Michigan moved by a country road to make connection between the First and Second brigades. Gregg's division followed Torbert as a reserve and support but did not become engaged.

Cold Harbor was a very important strategic point, as can be seen by a glance at the map, roads radiating from it in all directions. It was strongly held by Hampton's and Fitzhugh Lee's cavalry, supported by a brigade of infantry. They had thrown up breastworks of rails and logs, and made preparations for a stout resistance.

I reached the intersection of the country road with the left hand road before Devin appeared. My orders being to connect with him, I awaited his arrival, sending a few men out to keep watch in both directions. When Devin's advance came up, they saw these men and appeared to be suspicious of them, and did not advance very promptly. As soon as I could, I gave them to understand who we were and what we were there for. Devin then moved along the main road and the Sixth deployed through the woods until touch with its own brigade was obtained.

Cold Harbor Battlefield. The Battle of Cold Harbor was the final victory won by Robert E. Lee's army during the war. The battle caused a rise in anti-war sentiment in the Northern states, and it lowered the morale of the Union troops. But Lee had lost the initiative and was forced to devote his attention to the defense of Richmond and Petersburg. He spent the most of the remainder of the war defending Richmond behind a fortified trench line.

In the meantime, a hard fight was in progress. Torbert, not hearing from Devin, changed his plans and attacked the enemy's left flank with the Reserve Brigade and the First and Fifth Michigan. This was most skillfully and successfully done. The flanking movement was led by the First and Second United States and the Fifth Michigan, still under Captain Magoffin. The final blow was struck by Major Melvin Brewer with one battalion of the First Michigan, whose charge mounted at the critical moment decided the fate of the field. The enemy, who had been putting up a very hard fight, did not await this charge but threw down their arms and fled, the pursuit being followed up to a point a mile and a half beyond the town. The Sixth took little part except to fill the gap between Custer and Devin. The latter found the Confederate right flank too strong to circumvent, and added one more to the long list of lost opportunities.

Thus, Cold Harbor, the key to the maneuvers of the two armies, came into possession of the Union cavalry, but there was no infantry support within ten miles, the result having been unexpected by Meade, and Sheridan decided that it would not be safe for his command to try to hold it unsupported. He, however, notified the General of the Army what he had done, and withdrew his cavalry after dark to the position of the night before. Grant, realizing the importance of the capture, directed Sheridan to return and hold Cold Harbor at all hazards until the infantry could get up.

The march was retraced, and reaching the position before daylight, the breastworks which the enemy had thrown up were brought into service, strengthened as much as possible, and the division dismounted placed in line behind them. Ammunition boxes were distributed on the ground by the side of the men so they could load and fire with great rapidity. This was a strong line in single rank deployed thick along the barricade of rails. Behind the line only a few yards away were twelve pieces of artillery equally supplied with ammunition. The brigade was thus in readiness to make a desperate resistance to any attack that might be made.

The only mounted man on the line was General Custer, who rode back and forth giving his orders. The Sixth was lying down behind the rails and directly in front of the artillery, the pieces being so disposed as to fire over our heads. I do not remember any other engagement in which so many pieces of artillery were posted directly on a skirmish line with no line of battle behind it and no reserves. It was an expedient born of a desperate emergency.

In front of the line was open ground. Two hundred yards to the front were woods. In the woods, the Confederate infantry was in bivouac. Kershaw's division was in front of the Michigan Brigade. Before the first streaks of dawn began to appear in the east, their bugles sounded the reveille, and there was immediate commotion in the Confederate camps. So close to us were they that the commands of the officers could be heard distinctly.

Soon after daybreak an attack was made on the right of the line. As soon as the enemy emerged from the woods, General Custer ordered all the twelve pieces of artillery to fire with shell and canister, which they did most effectively. So furious was the fire that the Confederate infantry did not dare to come out of the woods in front of Custer's left where the Sixth was, the artillery and the fire from the Spencers from behind the rails keeping them back. An attempt was made to charge the part of the line where the First Michigan was posted, but each time it was repulsed. Here Captain Brevoort, one of the bravest and best officers in the brigade, was killed. Captain William M. Heazlett, another fine officer, was wounded. They both belonged to the First Michigan.

During the progress of the engagement, when the first attempt of Kershaw's infantry to come out of the woods had been repulsed, and there was a temporary suspension of the firing, General Custer, riding along the line in rear of the artillery, noticed that several of us who were lying down behind the barricade were directly in front of one of the brass pieces. Though these pieces were firing over our heads, they were very nearly, if not quite, on the same level as the barricades. He, with characteristic thoughtfulness, called my attention to the danger of remaining where we were, and I moved away from in front of the gun to a position in front of the interval between two of them, directing the others to do likewise.

The three men who were with me were Lieutenant William Creevy, Corporal John Yax and private Thomas W. Hill of troop C. Hill moved to the right when I moved to the left, but Creevy and Yax were slow about it. The very next time the gun was fired, there was a premature explosion, which killed Yax and wounded

Creevy. Hill was a boy only seventeen years of age, one of the recruits of 1863-64. He survived the war and is now cashier of the Cleveland National Bank of Cleveland, Ohio, and one of the most influential and respected businessmen of that city. Another one of those young recruits of 1863-64 was A.V. Cole, corporal in the same troop as Hill. He was badly wounded in the action at Haw's Shop, May 28th. For many years he was Adjutant General of the state of Nebraska.

This line was successfully held, a most meritorious performance, by the cavalry until nearly noon, when the Sixth corps came on the ground and relieved it. Never were reinforcements more cordially welcomed. Never did the uniform and arms of the infantry look better than when the advance of the Sixth Corps made its appearance at Old Cold Harbor. In solid array and with quick step, they marched out of the woods in rear of the line and took our places. The tension was relaxed, and for the first time since midnight the cavalryman drew a long breath.

This was the beginning of the intimate association of the First Cavalry Division with the Sixth Corps. So close a bond did it become that its hold was not released until the war closed. It was a bond of mutual help, mutual confidence and respect. The Greek cross and the cross sabers were found together on all the battlefields of the Shenandoah Valley, and we shall see how at Cedar Creek they unitedly made a mark for American valor and American discipline unexcelled in all the annals of war. There, side by side, Wright and Ricketts, Getty and Wheaton stood with Merritt and Custer in the face of an enemy flushed with success, and refused to be beaten until Sheridan came on the field to lead them to victory.

The division then moved back near Old Church and went into camp. June 2nd went into camp at Bottom's Bridge, where we remained skirmishing with cavalry across the river. June 6th found the First and Second Divisions in camp at Newcastle Ferry on the Pamunkey River, in readiness for what is known in the records and in history as the Trevilian Raid, conducted by General Sheridan in person.

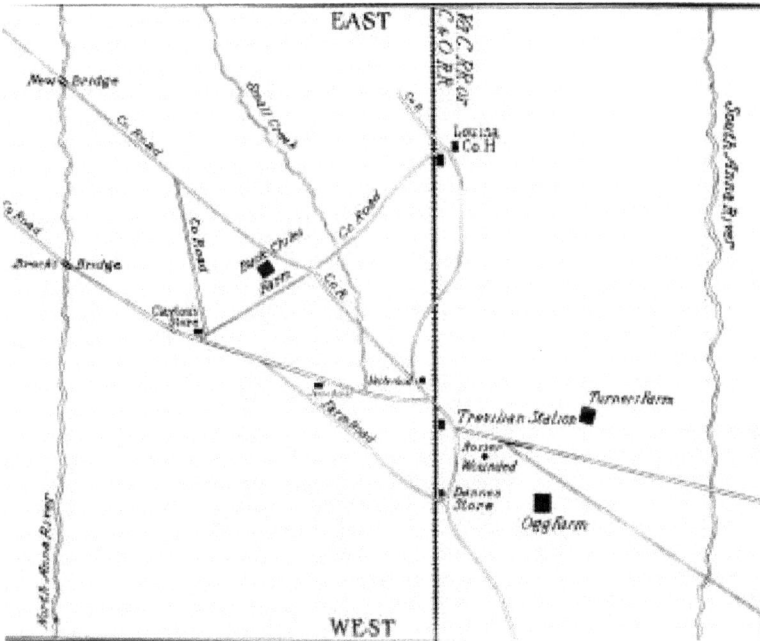

THE TREVILIAN RAID

THE contents of this chapter constitute the latest contribution of the author to the literature of the events recorded in this book. Much of that which has gone before and all of what follows was written many years ago. But in this final draft, every line has been revised. Time and the ripeness of years have tempered and mellowed prejudice; the hasty and sometimes intemperate generalizations of comparative youth have been corrected by more mature judgment; something of ill advised comment and crudity has been eliminated. Many of his conclusions and even the accuracy of some of his statements of fact, he realizes fully, may not remain unchallenged; yet it has been his honest endeavor and purpose to give, so far as in him lies, a truthful and impartial recital of those salient memories that remain to him of the stirring experiences of the youthful days, when as a boy he "followed the fortunes of the boy general" in the campaigns of 1863-64, in the great Civil War.

The outlines of the sketches herein made have been drawn from the official "records of the rebellion" which have been carefully consulted; the details for the most part have been taken from the storehouse of a somewhat retentive memory; something of color and atmosphere necessarily has been left to the imagination. It is a picture that he would present, rather than a dry recital of dates and places or a mere table of statistics. The importance of these things need not be lessened by seeking to give them an attractive form.

The writer must confess, also to an ambition to contribute something, albeit but a little, toward giving to the Michigan Cavalry Brigade the place in history which it richly earned; so that it may receive in its due proportions the credit which it deserves for the patriotic and valiant services rendered on so many battlefields. And especially does it seem to be to him a duty to do this for the regiment in which it was his privilege and good luck to serve. This ambition, however, was nearly stifled soon after its birth by an experience very galling to the pride of a well meaning, if sensitive and fallible historian.

It was something like twenty years ago that a paper on the Battle of Cedar Creek, prepared with conscientious care and scrupulous fidelity to the facts as the writer understood them, was mailed to General Wesley Merritt, with the request, couched in modest and courteous phrase, that he point out after having read it any inaccuracies of statement that he might make a note of, as the article was intended for publication.

The distinguished cavalry officer replied, in a style that was bland, that he had "long since ceased to read fiction;" that he no longer read "even the Century war articles;" that an officer one month would give his version of things which another officer in a subsequent number of the same magazine would stoutly contradict; and that he was heartily tired of the whole business.

General Merritt was however, good enough to give in detail his reasons for dissenting from the writer's account of a certain episode of the battle, and his letter lent emphasis to the discussion in one of the early chapters of this volume concerning men occupying different points of view in a battle. This particular matter will be more fully treated in its proper place. One must not be too sure of what he sees with his own eyes and hears with his own ears, unless he is backed by a cloud of witnesses.

Moreover, this was notice plain as holy writ that no mere amateur in the art of war may presume, without the fear of being discredited, to have known and observed that which did not at the time come within the scope of those who had a recognized status as professional soldiers, and find its way into their official reports. Indeed, a very high authority as good as told the writer in the war records office in Washington that no man's memory is as good as the published record, or entitled to any weight at all when not in entire harmony therewith.

It is evident that this rule, though perhaps a proper and necessary one to protect the literature of the war against imposition and fraud, may very easily bar out much that is valuable and well worth writing, if not indispensable to a fair and complete record, provided it can in some way be accredited and invested with the stamp of truth.

It was quite possible for brigade and even regimental commanders, not to draw the line finer still, to have experiences on the battlefield of which their immediate superiors were not cognizant; nor is it necessary to beg the question by arguing that all commanding officers were allowed to exercise a discretion of their own within certain limits.

Official reports were oftentimes but hastily and imperfectly sketched amidst the hurry and bustle of breaking camp; or on the eve of battle, when the mind might be occupied with other things of immediate and pressing importance. Sometimes they were prepared long afterwards, when it was as difficult to recall the exact sequence and order of events as it would be after the lapse of years. Some of the "youngsters" of those days failed to realize the value their reports would have in after years as the basis for making history. Others were so unfortunate as to have them "lost in transit," so that although they were duly and truly prepared and forwarded through the official channels, they never found their way into the printed record.

Attention already has been called to the absence of reports of the commanders of the Michigan Cavalry Brigade regiments for the Gettysburg campaign. General George B. Davis, U.S. Army, when in charge of the war records office in Washington, told the writer that he had noticed this want and wondered at it. He could not account for it. A like misfortune befell the same regiments when they participated in the Kilpatrick raid. Only a part of their reports covering the campaign of 1864, including the Trevilian Raid, were published. In this respect, the Sixth Michigan suffered more than either of the others. Not a single report of the operations of that regiment for that period appears in the record, though they were certainly made as required.

General Custer's reports cover that regiment, of course, as they do the others in the brigade, but it is unfortunate that these are not supplemented by those of the regimental commander. Until the volumes successively appeared, he was not aware of this defect; nor did he ever receive from any source an intimation of it, or have opportunity to supply the deficiency. Hence, it appeals to him as a duty to remedy, so far as it can be done at this late day, the omissions in the record as published of this gallant regiment.

From the beginning to the end of the campaign of 1864 in Virginia - from the Wilderness, May 4th, to Cedar Creek, October 19th - except for a single month when he was in command of the brigade, the writer was present with and commanded the Sixth Michigan Cavalry. Not a single day was he absent from duty, nor did he miss a battle or skirmish in which the regiment was engaged. Reports were made, but as we have shown, they did not find their way into the War Department. No copies were retained, so there is a hiatus in the record. There are numerous cases of a similar kind. Some officers, there is reason to believe, were smart enough to seek and were given the opportunity to restore the missing links.

The Trevilian Raid resulted from the seeming necessity of drawing the Confederate cavalry away from the front of the Army of the Potomac while the movement of the latter from the Chickahominy to the James was in progress. Sheridan was ordered to take two divisions and proceed to Charlottesville on the Virginia

Central Railroad. Incidentally, he was to unite there with the force operating under General Hunter in the direction of Lynchburg. He decided to take the First and Second divisions (Gregg and Torbert). Wilson with the Third Division was to remain with the army, taking his orders directly from General Meade.

As we have seen, the expeditionary force, before making the start, was at Newcastle Ferry on the south bank of the Pamunkey River. Three days' rations to last five days were ordered to be taken in haversacks; also two days' forage strapped to the pommels of the saddles; one hundred rounds of ammunition - forty on the person, sixty in wagons; one medical wagon and eight ambulances; Heaton's and Pennington's batteries; and a pontoon train of eight boats. The brigade commanders were Custer, Merritt, Devin, Davies and Irvin Gregg. In the Michigan Brigade, there had been some changes since Cold Harbor. Colonel Alger had returned and resumed command of his regiment. Major Melvin Brewer of the First Michigan had been promoted to lieutenant-colonel and assigned to command of the Seventh Michigan, his appointment dating June 6th.

There is a certain something about the events of that war that makes them stand out in bold relief, like architectural images on the façade of an edifice. They throw all other recollections of a lifetime into the shade. As I sit at my desk writing, with memory at elbow as a prompter, it is difficult to believe that today (May 7, 1908) it lacks but one short month of being forty-four years since those preparations were making on the banks of the Pamunkey River for a cavalry expedition in some respects more strenuous, more difficult than any which had preceded it. Yet those incidents are burned into the memory, and it seems that after all, it may have been but yesterday, so deep and lasting were the impressions then produced. As the well focused optical image is transferred to a sensitized surface, reproducing the picture, so were those scenes fixed in the mind with photographic certainty, to be retained as long as memory lasts, somewhat faded by time it may be, but complete in outline if not in details.

The campaign of the previous month had been a hard one for the cavalry. Aside from the fact that he was leaving one third of his force behind, Sheridan's corps had been decimated. A large number of his troopers had been killed and wounded, or rendered *hors de combat* in other ways. The horses had suffered terribly and many of them had been shot. So only about half the number of mounted men fit for duty that followed the colors of the cavalry corps out of the Wilderness, May 8th, marched across the Pamunkey on the pontoon bridge, June 6th. Readers who have followed this narrative through the preceding chapters will readily understand this.

Sheridan's plan[27] was to move along the north bank of the North Anna to a point opposite Trevilian Station, on the Virginia Central Railroad; then cross the North Anna by one of the bridges or fords, and by a rapid movement capture the station, destroy the railroad from Louisa Courthouse to Gordonsville, and proceed thence to Charlottesville, where the expected junction with Hunter was to be made. If this plan should succeed, the two forces thus united were to advance on Lynchburg and do what, as a matter of fact, Sheridan did not accomplish until the spring of 1865. Instead of marching to Charlottesville, Hunter went the other way, and that feature of the expedition was a failure. Breckinridge's corps of infantry was sent to Gordonsville, the Confederate cavalry succeeded in interposing between that place and Trevilian Station and Sheridan advanced no farther than the latter point.

Sheridan's march began on the morning of June 7th. Passing between the Pamunkey and the Mattapony rivers, he reached Polecat Station on the Richmond and Potomac (Fredericksburg) railroad the evening of June 8th, and encamped there for the night. The next day, the march was resumed, passing through Chilesburg to the North Anna, and along the bank of that river to Young's Mills, where the entire command bivouacked. June 10th, he journeyed to Twyman's Store and crossed the North Anna at

Carpenter's Ford near Miner's Bridge, between Brock's Bridge and New Bridge, encamping for the night on the road leading past Clayton's Store to Trevilian Station.

In the meantime, as soon as Sheridan's movement was discovered, two divisions of Confederate cavalry (Hampton's and Fitzhugh Lee's) under Hampton - the latter's division commanded by Butler - started by the direct road between the Annas for Gordonsville for the purpose of intercepting Sheridan. Breckinridge timed his movements to make his line of march parallel with that of Sheridan. Hampton, having the shorter distance to cover, although he started two days later than his adversary, was able to anticipate the latter in arriving and was between Gordonsville and Trevilian Station the night that Sheridan crossed the North Anna.

Fitzhugh Lee at the same time was near Louisa Courthouse, the two Confederate commanders thus being separated by a distance of some six or seven miles on the evening of June 10th. The Federal cavalry was all together and in position favorable for preventing a union of the Confederate forces by a sudden movement in the morning. Both commanders were looking for a battle on the following day and had made their plans accordingly.

Hampton had with him the three brigades of Rosser, Butler and Young; while the other division consisted of the brigades of Lomax and Wickham. It will thus be seen that while the Federal commander had a much smaller force than that which followed him on the raid of the previous month, his opponent was able to meet him with nearly twice the relative strength with which Stuart confronted him at Yellow Tavern. In other words, while Stuart fought him with the three brigades of Lomax, Wickham and Gordon (Hampton not being present) the latter at Trevilian Station had five brigades, including the big South Carolina brigade which fought so gallantly at Haw's Shop. More than that, Breckinridge's infantry was behind the cavalry, ready to reinforce it if needed.

Sheridan's camp was in the woods north of Clayton's Store,

and extending eastward as far as Buck Chiles's farm, Gregg on his left, Torbert on the right. His plan was to advance on Trevilian Station at an early hour on the morning of June 11th by the direct road from Clayton's Store. It was given to Gregg to look out for Fitzhugh Lee, who was expected to come into the action from the direction of Louisa Courthouse.

Hampton planned to advance from Trevilian Station with his own division and attack Sheridan at Clayton's Store. Lee was to take the road from Louisa Courthouse to the same point and form on Hampton's right. A glance at the map will show that the two roads intersect. Still another country road runs from Louisa Courthouse to Trevilian Station.

Sheridan formed his line of battle with Merritt on the right, Devin to Merritt's left, Custer and Gregg, en echelon, still farther to the left. Custer covered the road toward Louisa Courthouse. The Seventh Michigan picketed that road during the night. At a very early hour, the pickets of that regiment were attacked by Lee's advance. The First Michigan was sent to reinforce the Seventh. One brigade of Gregg's division was also sent out to meet Lee. The other one was formed on Devin's left. Sheridan then advanced and attacked Hampton instead of awaiting his attack.

Hampton moved from Trevilian Station with the two brigades of Butler and Young, Butler on the left. Rosser was sent to guard a road farther to the left, protecting that flank. Thus Rosser was isolated when the battle began, and Hampton came into action with but two brigades on the line. Fitzhugh Lee was headed off by the First and Seventh Michigan and Gregg's brigade, so that instead of coming to Hampton's assistance as intended, he was finally compelled to take the road leading directly to Trevilian Station instead of the one to Clayton's Store. It will be seen later that he arrived there at an opportune moment to prevent the complete destruction of Hampton's division.

The entire country between the North Anna River and the railroad was covered with timber and a dense undergrowth, except where there were occasional patches of cleared farmlands. When Torbert with his two brigades came into contact with Hampton, his line was found strongly posted in woods so dense that it was difficult to make headway against the defense. From the start however, Sheridan was the aggressor and Hampton was forced to fight a defensive battle.

In view of the rule laid down by General Sheridan himself, a criticism might be made on the tactics of the battle. But whether the error, if it was an error, should be laid at the door of the Chief of Cavalry or of General Torbert there is no way of finding out, though there is reason to believe that the former left the tactics on the field to be worked out by the division commanders. Custer was ordered to take a country road and pass around the flank to the rear of the enemy confronting Torbert. The exact location of this road was unknown, and Torbert states in his report that he was under a misapprehension about it; that it did not come out where he supposed it did; and that Custer by taking it lost touch with the other brigades which he was not able to regain until it was too late to accomplish the best results.

Such combinations rarely work out as expected, and Custer should have been put into action on the left of the line of battle; should have advanced with the division, keeping touch to the right, all the brigades in position to support each other. Then, by directing the entire movement in person, it is probable that Sheridan might have thrown his left forward, completely enveloping Hampton's right and crushing it before there was any possibility of receiving reinforcements. In that event, this turning movement would have been Custer's part of the battle, his regiments would have been kept together under his eye, and well in hand for a combined movement at the right moment. Complete success must have followed.

The road which Custer took leaves the North Anna River at New Bridge and runs to Trevilian Station. It crosses the Louisa Courthouse and Clayton Store Road east of Buck Chiles's farm. It intersects the direct road from Louisa Courthouse to Trevilian Station at a place designated on the map as "Netherland."

When Custer started out in the morning, the chances were that he would have a hard fight with Fitzhugh Lee at the outset. But it has been shown how by the interposition of the First and Seventh Michigan and one of Gregg's brigades, that officer was obliged to abandon the plan of reaching Clayton's Store and take the other road. So Custer, being relieved from pressure in that direction, started with the Fifth Michigan in advance, followed by Pennington's battery, to carry out his orders to get in Hampton's rear at or near Trevilian Station.

The advance guard was led by Major S.H. Hastings, one of the most daring officers in the brigade. At some point beyond the crossroads east of Buck Chiles's farm, the exact location being a matter of great uncertainty upon which the official reports shed no light whatever, Hastings discovered a train of wagons, caissons, led horses and other impedimenta, which he reported to the brigade commander and received orders to charge upon it, the charge to be supported by the entire regiment under Colonel Alger. This charge resulted in the capture of the outfit, but was continued for a long distance beyond the station, this being necessary in order to head off the train, which made a desperate effort to escape in the direction of Gordonsville. Custer's order to the Fifth did not contemplate continuing the pursuit beyond the station, since he was supposed to make a junction there with the other brigades of the First Division. But those two brigades were still fighting with Hampton, and the Fifth Michigan was directly in the latter's rear.

When this tumult arose in his rear, Hampton immediately recalled Rosser's brigade posted to protect his left flank, thereby leaving the way open for this foray around his right. Rosser, coming quickly upon the scene, not only intercepted Alger's retreat but proceeded to contest with the Fifth Michigan the possession of the captures which that regiment had made. But, I am outrunning my story:

The charge of the Fifth Michigan left Custer's front uncovered, and a force of Confederates which belonged to Young's brigade and had probably been looking out for Hampton's right flank and rear, threw itself across his path and boldly challenged his right to advance. This was not a large body of troops, probably the Seventh Georgia Cavalry, but it made up in audacity what it lacked in numbers. At that time - immediately after the charge of the Fifth Michigan - and before Rosser had begun his interference, Custer had with him only his staff and escort, and behind them was Pennington's battery which had no opportunity to come into action. The situation was apparently critical in the extreme.

The only available regiment at the time to throw into the breach was the Sixth Michigan, and that was just starting to move out of the woods where it had been encamped during the night. It was not supposed then that the battle was joined, and indeed the expectation was that the march was to be a continuation of that of the previous day, although the picket firing in the early morning indicated the close proximity of the enemy. But that had been the case for a morning or two before.

Before mounting, the officer in command had thoughtlessly acceded to the request of a brother officer to ride a spirited and nervous black horse belonging to the latter, as he expressed it, "To take the ginger out of him." In place of the regulation McClellan saddle, the horse was equipped with one of those small affairs used by jockeys in riding race horses. This had been picked up en route. Horse and saddle certainly made an attractive looking mount, but not such as one as a cavalry officer with a sound mind would select for close work on the battle line. The narration of these circumstances will enable the reader to judge of how little the subordinate officers knew of the real impending situation. It can be stated with absolute certainty that the officers of the Sixth were innocent of any knowledge of the fact that Custer had started out for a fight, up to the moment when they were ordered to mount and move out of the woods into a road running along the east side.

The commander of the regiment, mounted as described and leading the column of files, not having yet formed fours on account of the woods and brush, had barely reached the edge of the woods by the road when a member of the brigade staff brought the order to "Take the gallop and pass the battery." It is probable that this order was sent at the same time that the Fifth was sent forward to capture the train. Custer of course supposed that the Sixth was in column of fours in the road behind the battery. The commanding officer of the Sixth had moved out in compliance with orders and knew nothing about the conditions in front. The command, "Form fours, gallop, march" was given, and a touch of the spur sent the black steed flying toward the front, followed as quickly as possible by the leading squadron of the regiment. A regimental staff officer remained to repeat the order to the other squadrons as they came into the road successively.

Approaching the crossroads, the conditions were revealed as described in a previous paragraph. Custer and his escort were exchanging shots with their revolvers at short range with the Confederates in their front. The most remarkable coolness and courage were being displayed on both sides. The enemy certainly was commanded by an officer of resources who realized to the fullest extent the responsibility resting upon him to delay our further advance as long as possible. Custer never lost his nerve under any circumstances. He was however, unmistakably excited. "Charge them" was his laconic command; and it was repeated with emphasis.

Looking back to see that the leading squadron was pretty well closed up, I gave the command, "Draw sabers," and without waiting to form front into line or for the remainder of the regiment, the column of fours charged straight at the line of Confederates, the black horse leading. In a moment, we were through the line. Just how it was done is to this day more or less of a mystery. The enemy gave way - scattered to the right and left

- and did not await the contact. On down the road, one hundred, two hundred - it may have been five hundred yards, but not more than that, at breakneck speed, the charge continued. Then it was seen that there was no enemy in front of us. Where was the enemy?

Custer says in his report that Alger's orders were to stop at the station. The single word "charge" comprehended his order to me. Nothing was said about stopping. No warning was given that the Fifth had already charged and was ahead of us. Nor did I know it. The order had been obeyed to the letter. The enemy had apparently been dispersed. At all events, he had disappeared from our front. At such times, the mind acts quickly. The obvious course was to halt, rally, reform, see what was going on in rear, rejoin the brigade commander, get the regiment all together for work where we were most needed.

Finding that both hands were required to curb the excited steed, which up to that moment had not allowed another horse to come up with him, I returned my revolver to the holster, and when his speed began to slacken, Captain Vinton, commander of the charging squadron, came alongside, gave the command "Halt," which was twice repeated. My horse swerved to the right, and when brought to a standstill was a little way in the woods. The clatter of hoofs behind had told me that I was followed, and I supposed it was by my own troopers. Not so, however. Vinton either did not hear, or was too much under the influence of a pardonable excitement and zeal to heed the order to halt, and continued on down the road to and beyond the station, where he overtook the rear of the Fifth and proceeded to assist in the endeavor to bring away the captured property. He was attacked by Rosser, who made a lot of his men prisoners. The detachment that went with him did not rejoin the regiment until late in the afternoon and then less the men who had been captured.

The word "Surrender" uttered in imperious tones saluted my ear, and glancing over my left shoulder to find whence it came, I found that a well mounted and sturdy Confederate officer had come up from my left rear, and addressing me in language both profane and apparently designed to cast reflections on my ancestry, declared that if I did not comply instantly with his polite request he would complete the front cut on my head. His men circling around in front with their carbines in the position of "ready" seemed to hint that they considered his demand a reasonable one and expressed a purpose to assist in enforcing it.

Now, it is a maxim that no cavalry officer may surrender so long as he is not unhorsed. But in the situation in which I found myself, there did not seem to be an available alternative. I surrendered, gave up the black horse and the jockey saddle, and never saw either of them afterwards. After the experience described, I was glad to be rid of them on most any terms. Several others were captured at the same time and in the same way. One of them, after being dismounted, tried to run away but was quickly brought to a halt by a shot from a Confederate's gun which wounded him.

It appears that when we went through their line, the rascally Confederates rallied, and leaving Custer's front charged our rear. Custer says in his report that after "the Sixth Michigan charged, the rebels charged that regiment in rear." When he wrote that report, he had forgotten that it was only a portion - less than a third of the Sixth which charged. Two-thirds of the regiment was still back where he was and not yet in the action. There were two squadrons, one commanded by Captain Manning D. Birge, the other by Captain Don G. Lovell in reserve. In using the term squadron here, I mean what in the Civil War was known as a battalion (four troops). Vinton's squadron did not all take part in the charge.

Four Confederate cavalrymen undertook the duty of escorting myself and a young Sixth cavalryman who had been trapped in the same way to the rear through the woods. Anticipating that our attack would be followed up, we managed to delay our guards as much as possible, and had gone not more than a hundred yards when a yelling in the road proclaimed that the curtain had risen on the second scene of our little drama. Custer had ordered Birge to charge. Birge's advance put the Confederates to flight, what there were left of them.

The noise of the pursuit disconcerted our captors so that we took the chances and made our escape under cover of the thick undergrowth. They fired at us as we ran, but did not succeed in making a hit. Fortunately, Birge directed his course through the woods out of which the enemy had come and into which they had gone in their flight. In a minute, we met him coming with a squad of men. He was greatly rejoiced to find that he had rescued me from my disagreeable predicament, and looking back across the years, I can see and freely acknowledge that to no man on this earth am I under greater obligations than to Manning D. Birge. But for his approach, it might not have been possible for us to successfully make our break for freedom. That was the only time I ever was a prisoner of war, and then only for about ten minutes. Custer, referring to my capture, says that I was rescued by a charge of my own regiment led by Captain Birge.

Bidding Birge to follow my late captors, I hurried out to the road and thence to the crossroads from which we had started so short a time before. Custer was still there. His battery was there. Most of the Sixth was halted there. My recollection is that the First and Seventh about that time joined Custer after finding that Fitzhugh Lee had withdrawn from their front looking toward Louisa Courthouse. Birge's charge had cleared the road of the enemy, for the time being. Custer ordered that a rail barricade be thrown up across the road leading to the right, from which direction the attacks had been made on him. Putting the men of Vinton's and Birge's squadrons who were available at work, Lovell's squadron of four troops which was intact and well in hand under as good an officer as there was in the brigade, was posted in line mounted, parallel with the road and behind a screen of timber in readiness to repel any further attack.

Manning Birge

In a few minutes, Sergeant Avery, one of the men who had gone with Birge in pursuit of the enemy from whom I had escaped, came in with a Confederate prisoner splendidly mounted. Avery, with cocked revolver, was making his prisoner ride ahead of him and thus brought him in. Receiving orders to dismount, the man gave the horse a caress, and with something very like a tear in his eye said; "That is the best horse in the Seventh Georgia Cavalry."

The horse, with Avery's consent, was turned over to me to take the place of the captured black. He proved to be a prize. Handsome as a picture, kind and well broken, sound, spirited but tractable with a glossy coat of silky luster, he was a mount that a real cavalryman would become attached to and be proud of. I rode him, and he had the best of care until he succumbed to the cold weather and exposure near Winchester in the winter following. He was a finely bred southern horse and could not endure the climate.

Birge was not so fortunate. When he went after his prisoners, he caught a Tartar, or came very near it. The barricade was only partially completed, when yelling in front - that is in the road leading to the right - caused everyone to look in that direction. Birge and a few of his men were seen coming at full speed with what looked like a good big squadron of the enemy at their heels. Mounting the Seventh Georgia horse, I rode around the barricade and into the field where Lovell was with his battalion. He had been placed there for just such an emergency. Birge did not stop until he had leaped his mare over the barricade. When the Confederate column came up, Lovell surprised them with a volley right in their teeth, which sent them whirling back into the woods out of which they had come.

This was the end of the fighting at that point. Taking with him the Seventh under Lieutenant Colonel Brewer, and the battery, Custer then moved on toward Trevilian Station, leaving the First under Lieutenant Colonel Stagg and the Sixth to bring up and look out for the rear. The affray at the crossroads had occupied less time than it takes to tell it. In giving the story, it has been difficult to steer into the middle course between a seeming desire to give undue prominence to one's own part in the action on one hand, and affectation of undue modesty on the other. The only course appeared to be to narrate the incidents as they befell and leave it to the kind reader to judge the matter on its apparent merits.

When Custer approached the station, he found Rosser in his way on his front and right flank. Fitzhugh Lee, coming from Louisa Courthouse, also attacked his left flank. For a time, there was a melee which had no parallel in the annals of cavalry fighting in the Civil War, unless it may have been at Brandy Station or Buckland Mills. Custer's line was in the form of a circle and he was fighting an enterprising foe on either flank, and both front and rear. Fitzhugh Lee charged and captured a section of Pennington's battery. The Seventh Michigan led by Brewer recaptured it. Fragments of all the regiments in the brigade rallied around Custer for the mounted fighting, of which there was plenty, while the

First and Sixth dismounted took care of the rear. Custer was everywhere present giving directions to his subordinate commanders, and more than one mounted charge was participated in by him in person.

Torbert's attack with Merritt's and Devin's brigades was at length successful in routing Hampton, whose men were driven into and through Custer's lines. Many of them were made prisoners. An officer and twelve men belonging to the Seventh Georgia Cavalry, making for the rear as they supposed, came into the arms of the Sixth Michigan skirmishers at one time. The officer gave up his revolver to me, and it proved to be a very fine five shooting arm of English make.

In the final stages of the battle, Gregg concentrated against Fitzhugh Lee, Torbert effected his junction with Custer, and the latter was extricated from his difficult and dangerous predicament after performing prodigies of valor. The lines changed front and the Confederates were driven across the railroad, Hampton towards Gordonsville, Lee to the eastward. The two did not succeed in coming together that night, and Lee was obliged to make a wide detour in order to reunite with his chief on the afternoon of the next day, Sunday, June 12th.

The entire command encamped on the battlefield in the neighborhood of Trevilian Station for the night. The next morning, Gregg was set at work tearing up the railroad toward Louisa Courthouse. The First Division was given a rest until the afternoon when, at about three o'clock, although it was Sunday, the order came for the First Division to proceed in the direction of Gordonsville. In the meantime, the forces of Hampton and Lee had united, and as will be seen, had planned to stop Sheridan's further progress at all hazards. There is some reason to believe that a part of Breckinridge's infantry had come out from Gordonsville to reinforce Hampton. Such was the impression at the time, and one at least of Sheridan's commanders states in his report that he was confronted by infantry. The writer is of the opinion that the "infantry" was Butler's dismounted cavalry, which when in a good position as they were that day, could do as good fighting as any infantry in the Confederate service.

The Michigan Brigade moved out first and the Sixth had the advance. The order was to proceed to a certain point named and then halt until the division closed up. Memory does not recall what the place was, but is quite clear as to that being the specific direction given by General Custer to the officer in command of the advance regiment. We had gone but a short distance, not more than a mile or two at most, when the advance guard reported the enemy entrenched across the way. Skirmishing began at once between our mounted men in front and dismounted Confederates behind breastworks of considerable strength.

A squadron was deployed and Sergeant Avery was directed to make his way far enough into the woods to find, if possible, what we had in our front. He came back in about ten minutes and reported that the breastworks in our immediate front were thoroughly manned, and that he had seen a column of at least a thousand men moving into the entrenchments on the enemy's right, in front of our left flank. He was sent back to give Custer this information, and the general came up and ordered the entire regiment to be dismounted to fight on foot. The Sixth was put in on the right of the road, and directly thereafter the Seventh was sent in on the left. It did not take long to demonstrate that two regiments were not enough and the First and Fifth went into the action on the right of the Sixth. Then Torbert reinforced the line with the Reserve Brigade and a portion of the Second, all under Merritt. The entire division became engaged.

Several assaults were made upon the Confederate line, but without success. They were in each instance repulsed. Fitzhugh Lee got in on the right flank of the division and inflicted severe damage upon the Reserve Brigade. We have never been able to understand why, if it was intended to break the enemy's line, Gregg's division was not brought into the engagement to protect that flank. General Merritt in his report intimates that he had to do more than his share of fighting; that when the Reserve Brigade advanced to the assault on the right, it was supposed that the attack would be pressed on the left; that it was not so pressed and that his brigade suffered unduly on that account. This is another case of a man being unable to see all that is going on in a battle.

The Michigan Brigade was on the left of the line. It was the first brigade engaged. It began the fight and stayed in it until the end. Harder fighting has rarely been done than that which fell to the Michigan men in that battle. Several attempts were made to drive the enemy from their front. The First Michigan especially made a charge across an open field in the face of a terrible fire from behind breastworks, going half way across before they were repulsed. When the First Michigan could not stand before a storm of bullets, no other regiment in the cavalry corps need try. That is a certainty. The losses in killed and wounded were very severe, as will be shown in a table printed at the end of this chapter.

The fighting continued until ten o'clock that night, when Sheridan decided to withdraw and abandon the expedition. It is worthy of remark that the entire division was unable to advance one inch beyond the place where the advance guard first encountered the enemy and where Sergeant Avery made the reconnoissance which revealed to General Custer the true situation. Poor Avery was killed while doing his duty as he always did; in the very front of the battle in the place of greatest danger. Captain Lovell and Lieutenant Luther Canouse of the Sixth were wounded; Captain Carr, and Lieutenants Pulver and Warren of the First Michigan were killed, and Captain Duggan and Lieutenant Bullock of the same regiment wounded. Captains Hastings and Dodge of the Fifth were wounded; also Lieutenant Colonel Brewer of the Seventh was wounded on the eleventh.

The casualties in the two days' fighting at Trevilian Station were very severe. The losses in killed and died from wounds received in the action aggregated in the brigade forty one, as follows:[28]

First Michigan	13
Fifth Michigan	8
Sixth Michigan	17
Seventh Michigan	3
Total	41

Of prisoners lost there were in all two hundred and forty-two, distributed as follows:

First Michigan	39
Fifth Michigan	102
Sixth Michigan	58
Seventh Michigan	43
Total	242

Of those who were captured and held as prisoners of war, eighty-eight died in southern prisons - most of them in Andersonville - as follows:[29]

First Michigan	12
Fifth Michigan	35
Sixth Michigan	26
Seventh Michigan	15
Total	88

The Battle of Trevilian Station practically ended the fighting which was done by the Michigan Brigade in the campaign from the Rapidan to the James. Sheridan's retreat was skillfully conducted, but was not especially eventful. A tabulated statement of the losses in the command, beginning in the Wilderness, May 6th, and ending at Trevilian Station June 12th, is appended hereto. By losses, I mean killed in action or died of wounds received in action. It is not possible to give a reliable statement of the wounded, reports of regimental commanders being very deficient in that particular. The table is compiled from the official records in the office of the adjutant general of Michigan and is believed to be approximately correct:

	First Michigan	Fifth Michigan	Sixth Michigan	Seventh Michigan	Total
Wilderness	2	3	4	--	9
Todd's Tavern	3	--	--	1	4
Beaver Dam Station	1	--	--	--	1
Yellow Tavern	14	7	3	9	33
Meadow Bridge	--	--	2	--	2
Hanovertown	--	--	3	--	3
Haw's Shop	5	13	18	6	42
Old Church	2	--	--	1	3
Cold Harbor	5	1	1	3	10
Trevilian Station	13	8	17	3	41
Total	45	32	48	23	148

Recapitulation - Killed and died of wounds, the Rapidan to the James:

First Michigan	45
Fifth Michigan	32
Sixth Michigan	48
Seventh Michigan	23
Total	148

In General Merritt's official report[30] for the period May 26th to June 26th, he makes the following statement: "The losses in killed and wounded, (in the Reserve Brigade,) are annexed in tabular statement. As they number more than the loss of the entire rest of the command, they sufficiently attest the severe services of the brigade."

When General Merritt says "the entire rest of the command,"

we shall assume that he means "the entire rest" of the First Division. We have no desire to make invidious comparisons, and have avoided doing so throughout these recollections. The Reserve Brigade was a fine brigade and always fought well, and never better than at Trevilian Station and in the battles immediately preceding that engagement. To prove that his comparison was not warranted, it is necessary only to refer to the official records. On page 810 of the same volume,[31] appended to the report of General Torbert, for the same period covered by General Merritt's report, we find:

Reserve Brigade

Officers killed	6
Officers wounded	17
Officers killed and wounded	23

Reserve Brigade

Men killed	57
Men wounded	275
Total men killed and wounded	332
Total officers and men killed and wounded	355

Second Brigade

Officers killed	2
Officers wounded	15
Officers killed and wounded	17

Second Brigade

Men killed	42
Men wounded	163
Men killed and wounded	205
Total officers and men killed and wounded	222

First Brigade

Officers killed	3
Officers wounded	12
Officers killed and wounded	15

First Brigade

Men killed	62
Men wounded	192
Men killed and wounded	254
Total officers and men killed and wounded	269

Total killed and wounded First and Second Brigades	491
Total killed and wounded Reserve Brigade	355

The Reserve Brigade comprised five regiments, two of volunteers and three of regulars. The Michigan Brigade consisted of four regiments, of course all volunteers. One third of the losses in killed and wounded at Trevilian Station in the Reserve Brigade were in the single regiment, the First New York Dragoons. My authority for this is still the official records. See page 186 of the volume already quoted and referred to in the footnote. Close analysis therefore shows that there are inconsistencies in the official records, and unguarded statements in the official reports.

The rest of the month of June was consumed in the return march to the army. Owing to the necessity of caring for a large number of wounded and of guarding several hundred prisoners, to say nothing of an army of colored people of all ages and of both sexes who joined the procession, it was necessary to take a tortuous course which traversed the Spottsylvania battleground, touched at Bowling Green, followed the north bank of the Mattapony River, reaching King and Queen Courthouse June 18th.

From this place, the sick, wounded and prisoners were sent to West Point. On the 19th, we marched to Dunkirk on the

Mattapony River, which was crossed on a pontoon bridge and thence to the Pamunkey, opposite White House. June 21st, the entire command crossed the Pamunkey at White House and marched the next day (June 22nd) to Jones's Bridge on the Chickahominy. June 25th reached the James River and on the 28th crossed that river to Windmill Point. From here, the First and Second Divisions were sent to Reams's Station to the relief of the Third Division under Wilson, which had run into a situation similar to, if not more serious than that which Custer faced on the 11th at Trevilian.

King and Queen County Court House. Today, the Courthouse Tavern Museum includes the Old Fary's Tavern building as a turn-of-the-19th-Century tavern, including a restored dining lounge, parlor and two bedrooms. It is the oldest surviving building in the Courthouse Historic District.

Finding that officer safe, we returned to Lighthouse Point and settled down - after having fought and marched for fifty-six consecutive days - for a period of rest and recuperation. During the entire march from Trevilian to the James, Hampton hovered on the flank of Sheridan's column, watching for a favorable opportunity to inflict a blow but avoiding a general engagement. In crossing from the Pamunkey to the James, Sheridan was charged with the duty of escorting a train of 900 wagons from the White House to Douthat's Landing on the James. General Gregg was entrusted with the responsibility of protecting the right flank, which placed him in the post of danger, and the brunt of the fighting as well as the greater part of the honors of the movement fell to his share. Indeed, General Sheridan in his official report, written in New Orleans a year after the war closed, gave Gregg credit for saving the train.

The time from July 2nd, when we returned to Lighthouse Point on the James River, to July 26th, was quiet and uneventful. Many hundred convalescent wounded and sick men returned from hospital to duty; many also who had been dismounted by the exigencies of the campaign returned from dismounted camps. A fine lot of new horses were received. During the month, the condition of the animals was very much improved, good care and a plentiful supply of forage contributing to the result. The duty performed was to picket the left flank of the army, the Michigan regiments connecting with Crawford's Division of the Fifth Corps.

The story of the participation of the cavalry with the Second Corps in the movement to the north side of the James, which began on the forenoon of July 26th, has been so fully and so well told by General Sheridan in his reports and in his memoirs that nothing is left to be added. In fact there is little, if anything, in the part taken by any portion of the force taken across by Sheridan and Hancock to differentiate it from that played by the whole. The object of the movement was to draw the enemy's attention away from the lines around Petersburg preparatory for the explosion of the mine which was to take place on the 30th. In this, it was successful.

General Lee mistook the attack on his left for real instead of a feint, and detached enough troops to meet it to not only assure the success of the attack on Petersburg, if it had been made with determination, but to seriously menace the safety of the two corps engaged in the movement. General Sheridan truthfully says that "The movement to the north side of the James for the accomplishment of our part of the plan connected with the mine explosion was well executed, and every point made; but it was attended with such anxiety and sleeplessness as to prostrate almost every officer and man in the command."

This was the last incident of importance connected with the services of the First Cavalry Division with the Army of the Potomac in the year 1864. August 1st, Sheridan was ordered to the Shenandoah Valley, and selected the First and Third Cavalry Divisions to go with him.

Since this is in some sort a personal narrative, it may be of interest to mention that while at Lighthouse Point, I received my commission as colonel, and July 9th was mustered out of the United States service as major - with which rank I had been commanding the regiment - and was mustered in the new grade. The promotion, which was unsought, was due to a request made to the governor, signed by all the officers of the regiment serving in the field, and recommended by General Custer. On the original petition, on file in the adjutant general's office in Lansing, is an endorsement in the general's own handwriting.[32]

IN THE SHENANDOAH VALLEY

WHEN Grant sent Sheridan to take charge of things in the Shenandoah Valley and close that gateway to the north, he gave him one corps of infantry (Sixth) and two divisions of cavalry (First and Third) from the Army of the Potomac. The Michigan Cavalry Brigade, still commanded by General George A. Custer, was a part of that force. It embarked on transports at City Point, Virginia, August 3, 1864, and proceeded to Washington, D.C., thence by the way of Poolesville, Maryland, to Halltown, Virginia, in front of Harper's Ferry, arriving there August 10th, in time to join in the advance of the new army of the Middle Military Division,[33] under its new commander.

Gregg with the Second Division was left behind, under the immediate direction of General Meade, and thus, much to their regret, the Michigan men parted finally with that fine officer and his superb command, with whom they had been associated so intimately and honorably at Gettysburg, Haw's Shop and in many other places. When they rejoined the Army of the Potomac in the spring of 1865, he had retired from the service. They never saw him again, but from the eventful days of 1863 and 1864 to the present time, they have never ceased to respect him as a soldier and a man; and he always had their entire confidence as a commander of cavalry.

Sheridan wanted Early to cross into Maryland or to fight him in and around Winchester, but was in the dark as to his adversary's intentions or movements, so at daylight, August 11th, he started a reconnoissance in force. Custer led the way across the Opequon Creek toward Winchester, and soon ran into Early's infantry. A sharp fight followed, which showed that Early was retreating up the valley. Ransom's regular battery, attached to the brigade, was charged by Confederate infantry, which was met and repulsed by a countercharge of one battalion of the Sixth Michigan Cavalry led by Captain James Mathers, who was killed. Sheridan had left the gateway via the fords of the Potomac River open, but Early was too foxy to take the lure. He was getting away as fast as he could to a place of safety.

Harper's Ferry. A cavern just outside Harper's Ferry served as a hiding place for John Mosby's Confederate guerrillas throughout the war. The entry was covered by brush and rocks, and was marked by a high cliff to identify the hideout. Beneath a trapdoor lay a tunnel leading to a stairway deep underground.

The pursuit was instantly taken up and the next day (12th) found us up against infantry again at Fisher's Hill, between Cedar Creek and Strasburg, a position impregnable against direct assault. For three days, we remained face to face with Early's infantry, constantly so close as to draw their fire and keep them in their entrenchments.

Confederate prisoners taken at Fisher's Hill. During the battle, despite a strong defensive position, the Confederate army of General Jubal Early was defeated by the Union Army commanded by General Philip Sheridan.

On the 16th, we marched to Front Royal. Sheridan had information that a force of infantry and cavalry had been dispatched from Richmond to reinforce Early, and incidentally, to strike Sheridan in flank or rear if he could be caught napping. The force consisted of Kershaw's division of infantry and Fitzhugh Lee's division of cavalry, all commanded by General R.H. Anderson. The route by which they were supposed to be approaching was through Chester Gap and Front Royal. If they could have reached the Shenandoah River and effected a crossing undiscovered, a short march would have brought them to Newtown, directly in rear of our army.

Custer crossed and marched through Front Royal, but no enemy was found. He then recrossed and took position on commanding ground half a mile or so back from the river, and ordered the horses to be unsaddled and fed and the men to cook their dinner. Headquarters wagons were brought up, mess chests taken out, and we were just gathering around them to partake of a hastily prepared meal when Fitzhugh Lee's cavalry, which had stealthily approached the ford, charged across and made a dash at our pickets.

Major H.H. Vinton of the Sixth Michigan was in command of the picket line and promptly rallying on his reserves, he courageously met Lee's attack and checked it. That dinner was never eaten. Custer's bugler sounded "to horse." As if by magic, the men were in the saddle. Custer dashed out with his staff and ordered the Fifth Michigan forward, to be followed by the other regiments. I supposed he would charge in the direction of the ford, where Fitzhugh Lee's cavalry was still contending with the Sixth Michigan. He did nothing of the kind.

Moving diagonally to the left, he reached the crest overlooking the river just in time to surprise Kershaw in the act of crossing. The Fifth Michigan deployed into line in fine style and opened such a hot fire with their Spencers that the head of Kershaw's column was completely crushed. Every Confederate who was across was either killed or captured. Many of those who were in the water were drowned, and those on the other side were kept there. Just then, Devin's brigade came up and helped to drive the cavalry across the river. The prisoners, all infantry, numbered from three to five hundred.

This rencounter at Front Royal was one of the most brilliant affairs of the war, and it illustrated well the marvelous intuition with which General Custer often grasped the situation in an instant of time. He did not anticipate Kershaw's movement, or he would not have given the order to unsaddle. It was a surprise, but he was alert and equal to the emergency. He was as bold to act as his perceptions were keen, and the incident recalls the intrepidity with which he met Rosser in the Wilderness under somewhat similar circumstances. Had he charged the cavalry, Anderson would have effected a crossing, and in a very short time might have had the Michigan Brigade at such disadvantage that it would have required all of Custer's boldness and skill to extricate it.

Front Royal. After the battle at Front Royal, the victorious Confederate First Maryland took charge of prisoners from the beaten Union First Maryland regiment. Many men recognized each other as former friends and family.

Custer divined that the dash of Lee's advance was a mask for the infantry, and by a movement that would have done credit to Murat or Ney, caught Kershaw astride the river and trapped him completely. The behavior of the Fifth Michigan was never more superb. I do not believe that a single regiment on either side, at any time during the entire war, performed a more brilliant deed. Major Vinton and his detachment also earned especial praise by interrupting without aid the first onset of Fitzhugh Lee's advance. The First and Seventh Michigan supported the Fifth in a most gallant manner.

General Custer had a lock of hair shot away from his temple and Lieutenant Granger of his staff was killed. Lieutenant Lucius Carver of the Seventh also lost his life in the engagement. After this fight, it was found that Sheridan had begun a retrograde movement down the valley to take a defensive position in front of

Halltown. The brigade brought up the rear, the Sixth Michigan acting as rear guard.

From the 16th to the 25th of August, it was marching and countermarching, picketing, reconnoitering and skirmishing continually. Both armies were maneuvering for position and advantage. Anderson's reinforcement had joined Early, and with the esprit of the Army of Northern Virginia, was constantly pushing close up to our lines and harassing us. The Michigan Brigade was mostly engaged with infantry and did not once, I believe, come into contact with the Confederate cavalry. It was a lonesome day indeed when their mettle was not put to the proof in a skirmish with either Kershaw or Breckinridge.

But one incident occurred to break the monotony. A part of the Fifth Michigan sent out to destroy some buildings supposed to contain supplies was surprised by Mosby's command, and fifteen men were killed outright. They were caught in a field where escape was impossible, and shot without mercy. The Sixth was sent out to reinforce the Fifth, and we searched far and near for the dashing partisan, but did not succeed in coming up with him. He departed as swiftly as he came and made his escape to the mountains.

Sheridan had in his turn been reinforced by Wilson's division of cavalry (Third), and on the 25th, Torbert[34] was sent out with Merritt's and Wilson's divisions to hunt up Fitzhugh Lee, who was reported to have gone in the direction of the fords leading into Maryland. At or near Kearneysville, a small force of cavalry was encountered which was driven rapidly along the road toward Leetown. Nearing the latter place, the inevitable infantry was found, and it turned out to be Breckinridge's corps, going north along the Smithfield and Shepherdstown pike. Shepherdstown is on the Potomac River, opposite Sharpsburg and the Antietam battleground.

It never will be known what Breckinridge was intending to do, for he turned on Torbert and did not resume his journey. The collision was a complete surprise to both parties, but Early's design, whatever it may have been, was disarranged, the movement was discovered, and though the cavalry had rather the worst of it, the information gained was worth all it cost. If Early had been contemplating an invasion of Maryland, he relinquished the design and did not revive it.

Mosby's battalion. John Singleton Mosby (center, standing), nicknamed the "Grey Ghost," was a Confederate army cavalry battalion commander. His command, the 43rd Battalion, 1st Virginia Cavalry, known as *Mosby's Rangers* or *Mosby's Raiders*, was a partisan unit noted for its quick raids and ability to elude Union Army pursuers by blending in with local farmers and townsmen.

Torbert, finding that he had more than he could handle, fell back toward Halltown, leaving Custer with his brigade for a rear guard. Custer, coming to a piece of woods south of Shepherdstown, neither the enemy nor our own cavalry being in sight, halted and had his men dismount to rest, they having been in the saddle since early morning.

We were all sitting or lying down with bridle reins in hand, taking our ease with more or less dignity, when a small body of Confederate horse made its appearance in the direction of Shepherdstown. The brigade mounted and started in pursuit but had hardly been put in motion when a line of infantry suddenly appeared in the woods we were vacating and opened fire upon us. The Confederate horsemen were driven away by the First and Seventh, and when General Custer rallied his brigade to confront the new danger, he found that Breckinridge had intercepted his retreat in the direction the rest of the cavalry had gone, and was closing in with a line that threatened to envelop the brigade.

Ruins of bridge at Shepherdstown

In a few moments, the enemy's right and left flanks began to swing in towards the river and he found himself face to face with two alternatives; to cut his way through, or fall back and take the risky chance of fording the river with Breckinridge close at his heels. Of course there was no thought of surrender, and Custer was not much given to showing his heels. Torbert left Custer to shift for himself. So far as I ever was able to learn, he made no effort to save his plucky subordinate, and the report that the

Michigan Brigade had been captured was generally credited in and around Harper's Ferry.

Custer, with surprising coolness, put his brigade into line, the Sixth on the right, the First, Fifth and Seventh to the left of the Sixth, the battery in the center, with backs to the river and faces to the enemy, and presented so bold a front that the infantry did not charge, but moved up slowly, maneuvering to get around and obtain possession of the ford in rear. Custer had the men cheer, and dared them to come on. With characteristic audacity, he actually unlimbered his pieces and gave them a charge or two right in their teeth; then limbering to the rear, he took successive new positions and repeated the performance.

While holding one of these points, a squadron of the First New York Dragoons, of Devin's brigade, which also in some way had been separated from its command, was driven in from the right, and riding up to where I was, the commanding officer, Captain Brittain, saluted and said; "Colonel, I am cut off from my own regiment and wish to report to you for duty." "Form your men to the right," I said. "It looks as if your aid would be very acceptable." "I have no cartridges. We have shot them all away." "You have sabers." "Yes, and by ----, they are loaded," he retorted, as he brought his men front into line on the right. Captain Brittain survived the war and came to Michigan to live. He often has sent me kindly reminders of his remembrance of the circumstances as narrated above. For many years, he had a home in Wexford County, and I last heard of him as prospering on the Pacific coast.

At that moment, the thing had a critical look. We were inside a horseshoe of infantry, the extremities of which very nearly reached the river. We had to go through that line, or through the river, or surrender. Breckinridge's line was in plain sight, not a half mile away, in the open and moving up in splendid order. So far as I am informed, Custer was the only man in the command who knew that there was a ford and that we were making for it. The rest were screwing their courage up to the task of breaking through. I never have ceased to admire the nerve exhibited by Captain Brittain when I told him it looked as if that was what we would have to do. He was an excellent officer and belonged to an excellent regiment. "My sabers are loaded."

The greatest coolness was displayed by General Custer and his entire command. There was not a hint of weakness or fear in any quarter. The brigade at each falling back ployed from line into column, and deployed into line again as if on parade, with Breckinridge and his corps for the spectators. Every movement was at a walk. There was no haste - no confusion. Every officer was on his mettle and every man a hero. Presently, Custer finally withdrew his battery, then the regiments one at a time, and slipped away into Maryland before the enemy realized what he was doing. The delicate duty of bringing up the rear was entrusted to Colonel Alger with his own regiment and the Sixth. I was ordered to report to him. The battery crossed first, then the First and Seventh, the brigade staff and general commanding.

The two regiments stood in line, watching the enemy closing in closer and closer until this was accomplished. Then Colonel Alger told me to go. He followed leisurely, and as the Fifth and Sixth were marching up the Maryland bank, a line of Confederates came up on the other side, and so astounded were they to see how we had escaped from their grasp that some of them actually cheered, so I have been informed. They had been deceived by the audacity of Custer and his men in the first place, and by the cleverness with which they eluded capture in the second.

The Battle of Shepherdstown was the last in which Colonel Alger was engaged. While the brigade was lying in camp on the Maryland side awaiting orders, he was taken sick and was sent to hospital by order of the brigade surgeon. He was assigned to special duty by order of President Lincoln and did not rejoin. The esteem in which he was held by General Custer and the confidence which that officer reposed in him to the last moment of his service in the brigade is amply evidenced by the selection of him to lead the attack on Kershaw at Front Royal and to bring up the rear at Shepherdstown.

The coolness and ability of the officers and the intrepidity of the men in the Michigan Cavalry Brigade were never more thoroughly tested than in those two battles. Custer was the hero of both, and Alger was his right arm. At Meadow Bridge, at Yellow Tavern and in all the battles of that eventful campaign, wherever they were associated together, wherever the one wanted a man tried, true, trained and trustworthy, there he would put the other. No misunderstandings that arose later can alter the significance or break the force of these cold facts.

In the Battle of Shepherdstown, Captain Frederick Augustus Buhl of the First Michigan was mortally wounded, dying a few days later. He was a Detroit boy and a classmate of mine in Ann Arbor when the war broke out. I was deeply grieved at his death, as I had learned to love him like a brother. He was conspicuous for his gallantry in all the engagements in which he participated, especially at Front Royal and Shepherdstown.

For two days, the brigade was lost. For a time, the report of its capture was generally credited. That it escaped, no thanks were due to General Torbert, the Chief of Cavalry. It is not likely that he knew anything about what a predicament he had left Custer in. The latter was, as usual, equal to the emergency.

I must pass now rapidly over a period of nearly a month, devoted for the most part to reconnoitering and retreating, to the eve of the Battle of Winchester. September 18th, about 8 o'clock in the evening, I went to headquarters to consult Dr. Wooster, brigade surgeon, about the condition of my health. I was very feeble, unable to eat, my eyes and skin the color of certain newspapers during the Spanish-American war. The doctor told me I must go home and insisted on making out a certificate of disability on which I might obtain a leave of absence. General Custer and most of his staff were present. I recall the circumstances very well, for a conversation in which the general asked me confidentially certain questions was incautiously repeated by someone who was present and returned to vex me after many years. I returned to my own camp about nine or half past nine, much cast down over the doctor's diagnosis of my case.

I mention all this to show how secretly the preparations for the eventful next day had been made. Not a word was dropped during my long interview with the general and his staff to arouse the suspicion that the army was about to attack Early. Yet at midnight, orders were received to be ready to move at two o'clock in the morning. Before that hour, horses were in line saddled, the men ready to mount. My cook made a cup of tea and a slice of toast. I drank half of the tea but could not eat the toast. At three o'clock, I mounted my favorite saddle horse "Billy," and by order of General Custer, led my regiment in advance of the division, toward Locke's Ford on the Opequon Creek. Nothing was said, but everyone knew that the army was in motion and that great things were in store for us.

BATTLE FIELD
WINCHESTER VA.
Sept 19th 1864

We neared the ford about daylight. There was a faint hope that the enemy might be taken by surprise and the ford captured without resistance, as it was a difficult crossing when bravely defended. In this however, we were doomed to disappointment, for an alert foe was found awaiting the attack. Indeed, they must have known of the Federal approach. Halting an eighth of a mile back and out of sight, Custer directed me to dismount the regiment and move in column of fours through a ravine at right angles with the creek. This ravine ran out at the top, where it reached the edge of a plowed field. This field extended some 100 or 150 yards to the crest overlooking the ford. Along the crest were fences, outbuildings and the farm house. Thence, there was an abrupt descent to the bed of the Opequon Creek.

This side hill slope consisted of cleared fields divided by fences. The hill where the house and barns were also sloped off to the left. The road to the ford skirted the hill to the left until it reached the bank, then ran parallel with the creek to a point about on a line with the farm house, where it turned to the left, and crossing the stream, took a serpentine course up the opposite slope. This latter was wooded and dotted on both sides of the road with piles of rails behind which were posted infantry sharpshooters.

The leading files had barely reached the summit at the edge of the plowed ground when the enemy opened fire on the head of the column of fours, before the regiment had debouched. There was momentary confusion as the sharpshooters appeared to have the exact range. The regiment deployed forward into line under fire, and with General Custer by my side, we charged across the field to the crest. Custer was the only mounted man in the field. Reaching the houses and fences, the Sixth proceeded to try to make it as uncomfortable for the Confederates as they had been doing for us. General Custer had gone back to direct the movements of the other regiments which were still under cover in the rear.

The charge prostrated me. I succeeded in getting across the

field, cheered on by the gallant Custer who rode half way, but then fell down and for a minute or two could not stand on my feet. I suppose my pale face and weak condition made a very fair presentment of a colonel demoralized by fright. It was a case of complete physical exhaustion. While it is probably for the most part moral rather than physical courage that spurs men into battle, it is equally true that good health and a sound body are a good background for the display of moral courage. If any of my friends think that jaundice and an empty stomach are a good preparation for leading a charge across a plowed field in the face of an entrenched foe, I hope that they never may be called upon to put their belief to the proof.

Opequon Creek. Among the Confederate dead at the ensuing Winchester battle was Colonel George S. Patton, Sr. His grandson and namesake would become the famous general of World War II, George S. Patton, Jr.

Custer then sent orders to engage the enemy as briskly as possible, and directed the Twenty-fifth New York[35] followed by the Seventh Michigan, to take the ford mounted. The attempt was a failure however, for the head of the New York regiment, after passing the defile around the left, when it reached the crossing, instead of taking it, kept on and circling to the right came back to the point from which it started; thus in effect reversing the role of the French army which charged up a hill and then charged down again. The Seventh Michigan, having received orders to follow the other regiment, obeyed and did not see the mistake until too late to rectify it, much to the chagrin of that gallant officer, Lieutenant Colonel Brewer, who commanded it and who later in the day laid down his life.

The First Michigan was then ordered up to make the attempt. That regiment moved in column down the road to the foot of the hill at the left and halted. Two squadrons, commanded by Captain George R. Maxwell, an officer of the most undoubted courage, were detailed as an advance guard to lead the charge. Some minutes passed and the sharpshooters began to annoy the mounted men of the First. Major Howrigan of that regiment, thinking that the Sixth ought to occupy the attention of the enemy so completely as to shield his men from annoyance, galloped up to where I was and excitedly asked if we could not make it hotter for them.

"They are shooting my men off their horses," he shouted. As he halted to deliver this message, a bullet struck the saddlebag in rear of his left leg. Reaching back he unbuckled the strap, lifted the flap, and pulling out a cork inserted in the neck of what had been a glass flask, exclaimed; "Blankety blank their blank souls, they have broken my whisky bottle." Saying which, he wheeled and galloped back through a shower of whistling bullets.

General Custer then sent orders by a staff officer for the Sixth to advance dismounted and support the charge of the First. The Seventh was also brought up mounted to charge the ford at the same time. Preparations for this final attack were just about completed when it was discovered that the Confederates were leaving their cover and falling back. Lowell had effected a crossing at another ford and was threatening the flank of the force in our front. The Sixth moved forward with a cheer. All the regiments advanced to the attack simultaneously, and the crossing of the

Opequon was won. A sharp fight followed on the other side with Early's infantry in which a portion of the First Michigan led by the gallant Captain Maxwell made a most intrepid charge on infantry posted in the woods behind a rail fence.

The cavalry soon had the force opposed to it fleeing toward Winchester, but making a stand from time to time, so that it took from daylight in the morning until nearly three o'clock in the afternoon to cover the distance of three or four miles between the crossing of the Opequon and the outskirts of the town after which the battle has been named, though, perhaps, it is more correctly styled "The Battle of the Opequon." Breckinridge's infantry and Fitzhugh Lee's cavalry, the same gallant adversaries who hustled us over into Maryland in such lively fashion during the previous month, stood in the way and made vigorous efforts to stop our progress. It was a case of hunted turned hunter, and the Wolverines more than balanced the account charged up against Breckinridge for the affair at Shepherdstown, August 25th. To borrow an illustration from the Rugby game, the cavalry kept working around the end for gains until a touchdown and goal were scored at five o'clock in the afternoon.

The battle was fought along the Martinsburg Pike, the enemy being flanked or driven from one position to another until all the brigades of Merritt's and Averell's[36] divisions, which had been converging toward a common point, came together about a mile out of Winchester. As that place was approached, the signs and sounds of a great battle became startlingly distinct. The roar of artillery and the rattle of small arms saluted the ear. Within sight of the fortifications around that historic town, a duel was raging between the infantry of the two armies. The lines of blue and gray were in plain sight off to the left. Puffs of smoke and an angry roar told where the opposing batteries were planted. Dense masses of smoke enveloped the lines. From the heights to the front and right, cannon belched fire and destruction.

The Union cavalrymen were now all mounted. The Michigan

Brigade was on the left of the turnpike; to its left, the brigades of Devin and Lowell; on the right, Averell's division of two brigades - five brigades in all - each brigade in line of squadron columns, double ranks. This made a front of more than half a mile, three lines deep, of mounted men. That is to say, it was more than half a mile from Averell's right to Merritt's left.

At almost the same moment of time, the entire line emerged from the woods into the sunlight. A more enlivening and imposing spectacle never was seen. Guidons fluttered and sabers glistened. Officers vied with their men in gallantry and in zeal. Even the horses seemed to catch the inspiration of the scene and emulated the martial ardor of their riders. Then a left half wheel began the grand flanking movement which broke Early's left flank and won the battle.

When the Michigan Brigade came out of the woods, it found a line of Confederate horse behind a stone fence. This was the last stand that Fitzhugh Lee, who commanded Early's cavalry, attempted to make. Indeed it was here probably, that he received the wound which rendered him *hors de combat*. General Wickham succeeded him. In the stone fence, there were places where the stones had fallen or had been thrown down, making openings through which horses could pass one or at most two at a time. The Union cavalrymen made for these openings, not halting or hesitating for an instant. The fence was taken, and breaking through they put to flight the Confederate cavalrymen who did not stop until they found refuge behind their infantry lines.

The Union line was broken up too. The country for a mile was full of charging columns - regiments, troops, squads - the pursuit taking them in every direction where a mounted enemy could be seen. The cavalry disposed of, the infantry was next taken in hand. Early's lieutenants, finding their flank turned, changed front and tried hard to stem the tide of defeat. The brigade became badly scattered. Custer with a portion of it charged right up to a Confederate battery, but failed to get it, not having force enough at that point. The portion of the command with which I found myself followed Lee's cavalry for a long distance, when reaching the top of a slope over which they had gone in their retreat, we found ourselves face to face with a strong line of infantry which had changed front to receive us, and gave us a volley that filled the air with a swarm of bullets. This stopped the onset for the time in that part of the field, and the cavalry fell back behind the crest of the hill to reform, and to tell the truth, to get under cover, for the infantry fire was exceedingly hot. They were firing at just the right elevation to catch the horses, and there was danger that our cavalrymen would find themselves dismounted, through having their mounts killed.

As my horse swerved to the left, a bullet struck my right thigh, and peeling the skin off that, cut a deep gash through the saddle to the opening in the center. The saddle caused it to deflect upwards, or it would have gone through the other leg. At the moment, I supposed it had gone through the right leg. Meeting General Custer, I told him with some pride that I was wounded and needed a surgeon. Not finding one, I investigated for myself and found that it was one of those narrow escapes which a pious man might set down to the credit of providence or a miracle. The wound was not serious, and I proceeded to assist in rallying as many men of the regiment as possible to report to General Custer, who was preparing for what proved to be the final charge of the battle.

This was made upon a brigade of infantry which was still gallantly trying to make a stand toward Winchester and in front of a large stone house. The ground descended from Custer's position to that occupied by this infantry. Custer formed his men in line, and at the moment when the enemy began a movement to the rear, charged down upon them with a yell that could be heard above the din of the battle. In a brief time, he was in their midst. They threw down their arms and surrendered. Several hundred of them had retreated to the inside of the stone house. The house was surrounded and they were all made prisoners.

This charge, in which the Michigan Brigade captured more prisoners than it had men engaged, was for perhaps an eighth of a mile within range of the batteries on the heights around Winchester, and until it became dangerous to their own men, the artillery enfiladed our line.

A fragment of one of those shells struck my horse, "Billy," in the nose, taking out a chunk the size of my fist, and he carried the scar till the day of his death (in 1888). This last charge finished the battle. Early retreated through Winchester up the valley, and nothing was left but to pursue. Sheridan broke Early's left flank by the movement of the cavalry from his own right. It was the first time that proper use of this arm had been made in a great battle during the war. He was the only general of that war who knew how to make cavalry and infantry supplement each other in battle. Had the tactics of the battle been reversed - that is to say, if Sheridan had moved against Early's right flank instead of his left - nothing could have prevented the capture or destruction of Early's army, as his retreat would have been cut off. But the way to the south was left open, and Early escaped once more to Fisher's Hill, where he was found the next day with the remnant - a very respectable remnant - of his army.

It may be of interest to some of my medical friends to remark here in passing, that the Battle of Winchester cured my jaundice. After crossing the Opequon, I began to be ravenously hungry and begged and ate hardtack until there was some danger that the supply would be exhausted. The men soon saw the situation, and when one saw me approaching he would "present hardtack" without awaiting the order. So I went into the mounted part of the engagement with a full stomach and in more ways than one with a better stomach for a fight.

I regret that it is impossible to give a complete list of casualties in the brigade. In the appendix to this volume may be found a roll of honor of all those who were either killed or died of wounds received in battle.

Of the officers, Lieutenant Colonel Melvin Brewer was mortally wounded, the bullet which killed him coming from the stone house in which the Confederates had taken refuge. Colonel Brewer went out in the First, of which regiment he had risen to be a major. With that rank, he was assigned to command the Seventh, and only in the previous June had been promoted to lieutenant colonel. He was an officer modest as he was brave; cool and reliable on all occasions. Lieutenant Albert T. Jackson of the First, killed early in the action, was a young officer of much promise. Captain William O. North of the Fifth, who lost his life in the melee near Winchester, was also a most excellent officer. Captain A.S. Matthews of the First was wounded.

The casualties on the whole were not so numerous as in some other less historic engagements, most of them befalling in the attacks on infantry early and late in the day. Breckinridge's infantry seems to have fired low when resisting the mounted cavalry, for the havoc among horses was very great. I find by my official report made to the adjutant general at the time, that seven officers in the Sixth alone had their horses shot, and there is no reason to suppose that this record exceeded that of the other regiments.

For the next three days, the brigade was in front of infantry at Fisher's Hill, so close to their lines as to draw their fire and keep them in their entrenchments. On the 22nd, Torbert was sent to Milford in the Luray Valley, taking Wilson's and Merritt's divisions. His orders were to break through one of the passes in the Massanutten Mountains and come out in rear of Early's army when Crook's flanking movement on the other side would have driven the Confederates out of the strong position at Fisher's Hill.

Crook's attack was completely successful and Early was soon whirling up the valley again. Torbert made a fiasco of it. He allowed Wickham, who succeeded Fitzhugh Lee after the latter was wounded, with at most two small brigades, to hold him at bay, and withdrew without making any fight to speak of. I remember very well how the Michigan Brigade lay in a safe position in rear of the line listening to the firing, and was not ordered in at all. If Custer or Merritt had been in command, it would have been different.

When Sheridan found that Torbert had retreated, he gave him a very peremptory order to retrace his steps and try again. Custer, followed by Lowell, was sent to the front, and in the forenoon of the 24th, Wickham's troopers were scattered in flight and the way opened for Torbert to carry out his instructions. Even then, the march was leisurely and the two big divisions arrived in Newmarket on the 25th only to find that it was too late. Early had escaped again.

On the 26th at Harrisonburg, Custer assumed command of the Second Division in place of Averell, and I succeeded to the command of the brigade. On the same day, the brigade was ordered to Port Republic, and seeing a wagon train on the other side, the Sixth and Seventh were sent across the south fork of the Shenandoah River to attack it. It turned out to be Kershaw's division, which had been shuttle-cocked back and forth between Lee's army and the valley all summer and which, once more on the wing to reinforce Early, was just coming from Swift Run Gap.

The two regiments were driven back, but retired in good order and recrossed the river. Sheridan then withdrew to Cross Keys, hoping to lure Early to that point, but was unsuccessful. The next day, Port Republic was reoccupied and the brigade established a picket line extended thence to Conrad's Ferry, a distance of twenty miles.

While occupying this position, the discovery was made that there were several good grist mills along the river that were also well stored with grist. There were plenty of men in the brigade who were practical millers, and putting them in charge, I had all the mills running very early in the morning, grinding flour and meal which the commissaries were proceeding to issue to the several regiments according to their needs, and we all flattered ourselves that we were doing a fine stroke of business. This complacent state of mind was rudely disturbed when about seven o'clock (the mills had been running some two hours or more), General Merritt, accompanied by his staff, dashed up, and in an angry mood which he did not attempt to conceal, began to reprimand me because the mills had not been set on fire.

The fiat had gone forth from General Grant himself, that everything in the valley that might contribute to the support of the army must be destroyed before the country was abandoned. Sheridan had already decided on another retrograde movement down the valley, and it was his purpose to leave a trail of fire behind, obeying to the letter the injunction of the General in Chief to starve out any crow that would hereafter have the temerity to fly over the Shenandoah Valley. The order had gone out the day before and the work was to begin that morning. Custer was to take the west and Merritt the east side and burn all barns, mills, haystacks, etc., within a certain area.

Merritt was provoked. He pointed to the west, and one could have made a chart of Custer's trail by the columns of black smoke which marked it. The general was manifestly fretting lest Custer should appear to outdo him in zeal in obeying orders, and blamed me as his responsible subordinate for the delay. I told him, with an appearance of humility that I am sure was unfeigned, that those mills would never grind again after what had passed. The wheels were not stopped but the torch was applied and the crackling of flames intermingled with the rumbling of the stones made a mournful requiem as the old mills went up in smoke and General Merritt's loyalty was vindicated.

It was a disagreeable business and - we can be frank now - I did not relish it. One incident made a lasting impression on the mind of every man who was there. The mill in the little hamlet of Port Republic contained the means of livelihood - the food of the women and children whom the exigencies of war had bereft of their natural providers, and when they found that it was the intention to destroy that on which their very existence seemed to depend, their appeals to be permitted to have some of the flour before the mill was burned were heartrending. Worse than all else, in spite of the most urgent precautions enjoined upon the officers in charge, the flames extended. The mill stood in the midst of a group of wooden houses and some of them took fire.

Mill on Santa Anna. Today, the North Anna Battlefield Park, opened in 1996 and maintained by Hanover County, Virginia, preserves a small section of the battlefield. Walking trails are available to view portions of the left side of the "inverted V" Confederate line up to Ox Ford.

Seeing the danger, I rode across and ordered every man to fall in and assist in preventing the further spread of the flames, an effort which was happily successful. What I saw there is burned into my memory. Women with children in their arms stood in the street and gazed frantically upon the threatened ruin of their homes, while the tears rained down their cheeks. The anguish pictured in their faces would have melted any heart not seared by the horrors and necessities of war. It was too much for me, and at the first moment that duty would permit, I hurried away from the scene. General Merritt did not see these things, nor did General Sheridan, much less General Grant.

The army began to fall back on the 6th of October, the cavalry bringing up the rear as usual, Merritt on the valley pike,

Custer by the back road along the east slope of the Little North Mountain. The work of incineration was continued and clouds of smoke marked the passage of the Federal army. Lomax with one division of cavalry followed Merritt, while Rosser with two brigades took up the pursuit of Custer on the back road. The pursuit was rather tame for a couple of days, but the sight of the destruction going on must have exasperated the Confederate troopers, many of whom were on their native heath, and put them in a fighting mood, for on the 8th they began to grow aggressive and worried the life out of our rear guard.

The Michigan Brigade had the rear. The Seventh was sent ahead to see that nothing escaped that came within the scope of Grant's order; the Fifth acted as rear guard; the First and Sixth in position to support the Fifth if needed. The pike formed the main street of the little town of Woodstock, the houses coming close to it on either side. On nearing that place, it was found that a fire started in some small barns and haystacks in the outskirts had caught in the adjoining buildings, and the town was in flames. Dismounting the two regiments and sending the lead horses beyond the village, orders were given to have the fires put out.

The men went to work with a will, but were interrupted in their laudable purpose by Lomax, who charged the rear guard into the town, and there was some lively hustling to get to the horses in time. The brigade was then formed in line in a good position facing Woodstock and awaited, indeed invited attack by the Confederates. Lomax, however, kept at a respectful distance until the march was resumed, when he took up the pursuit again. Thus it went, alternately halting, forming and facing to the rear, and falling back, until Tom's Brook was reached late in the afternoon. Then General Merritt directed me to send one regiment to reinforce Custer, who was being hard pressed by Rosser on the back road, and take the others and drive Lomax back.

The Seventh was sent to Custer and the First, Fifth and Sixth, the Sixth leading, drove the cavalry that had been annoying our rear at a jump back to Woodstock, a distance of about six miles. By that time, Lomax had his entire division up, and when we started to fall back again, gave us a Roland for our Oliver, following sharply, but always declining the invitation to come on when we halted and faced him. It was particularly annoying to the Fifth which brought up the rear and distinguished itself greatly by the stubborn resistance which it offered to the attacks of the enemy. Captain Shier's squadron of the First supported the Fifth with much spirit.

On the morning of the 9th, Sheridan told Torbert to go out and whip the cavalry that was following us or get whipped himself. It was a short job, and the Battle of Tom's Brook is regarded as one of the humorous incidents of the war. With slight loss in a very brief engagement, Rosser and Lomax were both routed, and the pursuit of the latter on the pike was continued for about twenty miles. The battle known in history as that of "Tom's Brook," was facetiously christened "The Woodstock Races," and the Confederate cavalry cut little figure in Virginia afterwards. The Michigan Brigade had a prominent part in the battle, being in the center and forming the connecting link between the First and Third Divisions.

In the opening attack, the Confederate center was pierced by the mounted charge of the Fifth, Sixth and Seventh Michigan, assisted by the Twenty-fifth New York. The First, being on picket during the previous night, had not returned to the command. I believe I am right in claiming that the first impression made on the enemy's line of battle was by these regiments, though the line was rather thin, for the reason that the heaviest part of Rosser's force had been massed in front of Custer and on the pike, making the center an especially vulnerable point.

When the flight began, they took to the roads, and the Michigan men being in the woods did not get very far into the horse race, as it was called. The First, coming from the picket line, trailed the leaders along the pike and managed to get a good deal of sport out of it with very little danger. I must now pass over the few intervening days to the crowning glory of the campaign, The Battle of Cedar Creek.

THE BATTLE OF CEDAR CREEK

THE engagement which took place in the Shenandoah Valley of Virginia, on the nineteenth day of October, 1864, will take its place high up in the list of the decisive battles of history. Like Blenheim and Balaklava, Cedar Creek will be remembered while literature lasts. One of its dramatic incidents furnished the theme for the poet's song, and *"Sheridan's Ride,"* like Horatius, will remain until the imagination can no longer be thrilled by the recital of the record of heroic deeds. Thus doth poesy erect monuments, more enduring than bronze or marble, to the memory of the brave.

Yet, the events of that day have been greatly misconceived.[37] The imagination, inflamed by the heroic verse of Read, and unaided by the remembrance of actual personal experiences in the battle, sees only the salient points - Gordon's stealthy march along the Massanutten Mountain; the Union troops in fancied security, sleeping in their tents; the absence of their great leader; the morning surprise; the rout; the mass of fleeing fugitives; the victors in exultant pursuit; Sheridan's ride from Winchester; the magic influence of his arrival on the field in arresting the headlong flight of the panic stricken mob; the rally; the reflux tide of enthusiasm; the charge back into the old camps; the glorious victory that succeeded humiliating defeat.

With all due allowance for poetical license, the conception of this battle which long ago became fixed in the public mind does a cruel injustice to the gallant men who were maimed or killed on that hard fought field. Enveloped in the mists of receding years; obscured by the glamour of poetry; belied by the vivid imagination of stragglers and camp followers who on the first note of danger made a frantic rush for Winchester, seeking to palliate their own misconduct by spreading exaggerated reports of disaster, the Union army that confronted Early at Cedar Creek, for many years made a sorry picture, which the aureole of glory that surrounded its central figure made all the more humiliating. It is due to truth and justice that every detail of that famous fight should be told, to the end that no undeserved shadow may rest upon the fame of the men and officers who took part in it - no unjust stain upon their record.

History, so called, has been misleading. It is true that Sheridan's narrative sheds much new light upon his part in the battle, and General Merritt, one of the leading actors, wrote a paper upon it for the Century series, though I doubt if it has been generally read, or if read, effective in modifying preconceived notions. An idea of that which has been written in the name of history may be gained from an extract taken from the American cyclopedia (vol. xvi) which says; "He (Sheridan) met the fugitives a mile and a half from town, (Winchester), and with a brigade which had been left in Winchester, moved upon the enemy, who had begun to entrench themselves."

The absurdity of such "history" ought to be self evident. Imagine if you can, a brigade of infantry following Sheridan on his wild ride of twenty miles, and then rushing to attack an army which, according to the tradition of which I have spoken, had just whipped four army corps. Of course, the statement is an absurd one. No brigade came from Winchester. No brigade could have come from Winchester; and had such a thing been possible, it would have constituted but a slight factor in the contest.

There were in the Federal army on that eventful morning, seven brigades of infantry (the Sixth corps) seven brigades of cavalry, not to mention one division (Grover's) of the Nineteenth

Corps, (four brigades), making eighteen brigades in all, that were neither surprised in their camps nor in the slightest degree demoralized at any time during the progress of the battle; and which had forced Early to stop short in his headlong career of victory long before the famous black charger brought his fiery rider to the field. The Eighth Corps which was surprised was a small corps of only five brigades, and although after Kershaw's onset, conducted by General Early in person, it was practically eliminated, there was a fine army left, which crippled as it was, was fully equal to the task of retrieving the disaster, and which as the event proved, needed only the guiding hand of Sheridan to put it in motion and lead it to victory.

It is not, however, the purpose of this paper to give all the details of that great battle, but to narrate what a single actor in it saw; to make a note in passing of some things that do not appear in the official records, that are not a part of the written history of the war; some incidents that are important only as they throw light on that which is bathed in shadow, though having for one of Custer's troopers an interest in themselves; to do justice to the splendid courage displayed by the cavalry, especially the Michigan cavalry on that occasion; to pay a tribute of admiration to the gallantry and steadfastness of the old Sixth Corps; and to the courage and capacity of the gallant Colonel Lowell, who was killed.

Cedar Creek is a small stream that rises in the Blue Ridge, runs across the valley, at that point but four miles wide, and pours its waters into the Shenandoah near Strasburg. It is very crooked, fordable, but with steep banks difficult for artillery or wagons, except where a way has been carved out at the fords. It runs in a southeasterly course, so that its mouth is four miles or more south of a line drawn due east from the point where it deserts the foothills on the west side of the valley. The valley itself is shut in between the Blue Mountains on one side and the Massanutten, a spur of the Great North Mountain, on the other. It is traversed, from north to south, by a turnpike road, a little to the left of the center, which road crosses Cedar Creek between Middletown and Strasburg.

Cedar Creek. Jubal Early's Confederates surprised Union troops near Cedar Creek and drove three Union Corps from the field. As Early paused to reorganize, Union General Phil Sheridan arrived from Winchester just in time to rally his troops and launch a crushing counterattack. Sheridan's victory at Cedar Creek ended any hope of further Confederate offensives in the Shenandoah Valley.

On the night of October 18th, 1864, the Federal army was encamped on the left bank of Cedar Creek, Crook's Eighth corps on the left flank, east of the pike and nearly in front of Middletown; Emory's Nineteenth corps to the right and rear of Crook and west of the pike; then, successively, each farther to the right and rear, the Sixth Corps, temporarily commanded by General James B. Ricketts; Devin's and Lowell's brigades of Merritt's (First) Cavalry Division; the Michigan Cavalry Brigade; and last but not least, Custer with the Third Cavalry Division. All faced toward the south, though posted en echelon, so that though Crook was some three or four miles south of Middletown, a line drawn due east from Custer's camp, intersected the pike a little north of that place.

For this reason, Early's flanking movement, being from the left through the camp of Crook, could not strike the flank of the other corps successively without shifting the line of attack to the north, while the Sixth Corps and the cavalry were able to confront his troops after their first partial success by simply moving to the left, taking the most direct route to the turnpike. The position which the Michigan cavalry occupied was somewhat isolated. Although belonging to the First Division, it was posted nearer the camp of the Third.

The brigade consisted of the four Michigan regiments and Captain Martin's Sixth New York independent horse battery. The First Michigan was commanded by Major A.W. Duggan, a gallant officer who was wounded at Gettysburg; the Fifth by Major S.H. Hastings; the Sixth by Major Charles W. Deane; the Seventh by Lieutenant Colonel George G. Briggs, the latter officer having only just been promoted to that position. The New York battery had been with us but a short time, but Captain Martin and his lieutenants ranked among the best artillery officers in the service.

For a few days only, I had been in command of the brigade. General Custer, who had led it from the time he was made a brigadier in June, 1863, was promoted to the command of the Third Division, and hastily summoning me went away, taking his staff and colors with him. I was obliged while yet on the march to form a staff of officers as inexperienced as myself. It was an unsought and an unwelcome responsibility.

For two or three days before the battle, our duty had been to guard a ford of Cedar Creek. One regiment was kept constantly on duty near the ford. The line of videttes was thrown out across the stream, connecting on the left with the infantry picket line and on the right with Custer's cavalry pickets. The Seventh Michigan was on duty the night of October 18th, the brigade camp back about a mile from the ford.

No intimation of expected danger had been received - no injunction to be more than usually alert. It was the habit of the cavalry, which had so much outpost duty to perform, to be always ready, and cavalry officers were rarely taken by surprise. Early's precautions had been carefully taken and no hint of his purpose reached the Union headquarters, and no warning of any immediate or more than usually pressing danger was given to the army. But somehow, I had a vague feeling of uneasiness that would not be shaken off. I believe now and have believed for many years that there was in my mind a distinct presentiment of the coming storm. I could not sleep, and at eleven o'clock was still walking about outside the tents.

It was a perfect night, bright and clear. The moon was full, the air crisp and transparent. A more serene and peaceful scene could not be imagined. The spirit of tranquility seemed to have settled down at last upon the troubled Shenandoah. Far away to the left lay the army, wrapped in slumber. To the right, the outlines of the Blue Mountains stood out against the sky and cast dark shadows athwart the valley. Three-quarters of a mile away, the white tents of Custer's camp looked like weird specters in the moonlight. Scarcely a sound was heard. A solemn stillness reigned, broken only by the tread of the single sentry, pacing his beat in front of headquarters. Inside, the staff and brigade escort were sleeping. Finally, a little before midnight, I turned in, telling the guard to awaken me at once should there be firing in front and to so instruct the relief.

I cannot give the exact time; it may be I did not know it at the time; but it was before daylight that the sentinel awoke me. Not having undressed, I was out in an instant and listening, heard scattering shots. They were not many, but enough to impel me to a quick resolve. Rousing the nearest staff officer, I bade him have the command ready to move at a moment's notice.

In an incredibly short space of time, the order was executed. The tents were struck, the artillery horses attached to the gun carriages and caissons and the cavalry horses saddled. No bugle call was sounded. The firing grew heavier, and from the hill where

Custer was, rang out on the air the shrill notes of Foght's bugle, telling us that our old commander had taken the alarm. Rosser had attacked the pickets at the fords and was driving them in. He had done the same on one or two mornings before, but there was an unwonted vigor about this attack that boded mischief.

Cavalry Headquarters of Charles Lowell. Lowell commanded a brigade of cavalry in General Wesley Merritt's division and was mortally wounded during the Union counterattack at the Battle of Cedar Creek on October 19, 1864. Lowell was nominated as a brigadier general two days after his death.

The Federal cavalry had, however, recovered from their earlier habit of being away from home when Rosser called. They were always in and ready and willing to give him a warm reception. He found that morning that both Merritt and Custer were "at home." In a moment, a staff officer from General Merritt dashed up with orders to take the entire brigade to the support of the picket line. Moving out rapidly, we were soon on the ground. The Seventh Michigan had made a gallant stand alone, and when the brigade arrived, the enemy did not see fit to press the attack, but contented himself with throwing a few shells from the opposite bank, which annoyed us so little that Martin did not unlimber his guns.

A heavy fog had by this time settled down upon the valley. The first streaks of dawn began to appear, and it soon became evident that the cavalry attack upon the right flank was but a feint and that the real danger was in another quarter. Far away to the left, for some time, volleys of musketry had been heard. With the roll of musketry was intermingled, at intervals, the boom of cannon, telling to the practiced ear the story of a general engagement. The sounds increased in volume and in violence, and it was no difficult matter to see that the Union forces were falling back, for farther and farther to the left and rear were heard the ominous sounds. From the position we occupied, no infantry line of battle was to be seen.

Soon after the Michigan Brigade had taken its position at the front, Colonel Charles R. Lowell rode up at the head of the Reserve Brigade. Colonel Lowell was a young man, not much past his majority, and looked like a boy. He was a relative of James Russell Lowell, and had won distinction as colonel of the Second Massachusetts Cavalry. He had succeeded Merritt as commander of the Reserve Brigade. He had a frank, open face, a manly, soldierly bearing, and a courage that was never called in question. He was a graduate of Harvard, not of West Point, though he had been a captain in the Sixth United States Cavalry.

Colonel Lowell informed me that his orders were to support the Michigan men if they needed support. No help was needed at that time. I told him so. The enemy had been easily checked, and at the moment, had become so quiet as to give rise to the suspicion that he had withdrawn from our front, as indeed he had. A great battle was raging to the left, and in response to the suggestion that the army seemed to be retreating, he replied; "I think so," and after a few moments reflection said; "I shall return" and immediately began the countermarch. I said to him; "Colonel, what would you do if you were in my place?"

"I think you ought to go too" he replied, and presently

turning in his saddle, continued; "Yes, I will take the responsibility to give you the order," whereat, the two brigades took up the march toward the point where the battle, judging from the sound, seemed to be in progress. How little either of us realized that Lowell was marching to his death. It was into the thickest of the fight that he led the way, Michigan willingly following.[38]

A startling sight presented itself as the long cavalry column came out into the open country overlooking the battleground. Guided by the sound, a direction had been taken that would bring us to the pike as directly as possible, and at the same time approach the Union lines from the rear. This brought us out on a commanding ridge north of Middletown. This ridge as it appears to a participant looking at it from memory runs to and across the pike. The ground descends to the south a half mile or more, then gradually rises again to another ridge about on a line with Middletown. The Confederate forces were on the last named ridge, along which their batteries were planted, and their lines of infantry could be seen distinctly. Memory may have lost something of the details of the picture, but the outlines remain as vivid, now as then. The valley between was uneven, with spots of timber here and there and broken into patches by fences, some of them of stone.

The full scope of the calamity which had befallen our arms burst suddenly into view. The whole battlefield was in sight. The valley and intervening slopes, the fields and woods, were alive with infantry moving singly and in squads. Some entire regiments were hurrying to the rear, while the Confederate artillery was raining shot and shell and spherical case among them to accelerate their speed. Some of the enemy's batteries were the very ones just captured from us.

It did not look like a frightened or panic stricken army, but like a disorganized mass that had simply lost the power of cohesion. A line of cavalry skirmishers[39] formed across the country was making ineffectual efforts to stop the stream of fugitives who had stolidly and stubbornly set their faces to the rear. Dazed by the surprise in their camps, they acted like men who had forfeited their self-respect. They were chagrined, mortified, mad at their officers and themselves - demoralized; but after all, more to be pitied than blamed.

Cedar Creek Battlefield. Today, more than 1,450 acres of the Cedar Creek Battlefield are preserved as part of the Cedar Creek and Belle Grove National Historical Park.

But all these thousands, hurrying from the field, were not the entire army. They were the Eighth Corps and a part of the Nineteenth only, a fraction of the army. There, between ourselves and the enemy - between the fugitives and the enemy - was a long line of blue, facing to the front, bravely battling to stem the tide of defeat. How grandly they stood to their work. Neither shot nor shell nor volleys of musketry could break them. It was the old Sixth Corps - the "ironsides" from the Potomac army, who learned how to fight under brave John Sedgwick. Slowly, in perfect order, the veterans of the Wilderness and Spottsylvania were falling back, contesting every inch of the way. One position was surrendered only to take another. There was no wavering, no falling out of ranks, except of those who were shot down. The next morning, one passing over the ground where those heroes fought could see where they successively stood and breasted the storm by the dead men who lay in line where they had fallen. There were two or three lines of these dead skirmishers. The official record shows that the Sixth Corps on that day lost 255 men killed and 1600 wounded.

The two brigades had reached a point where the entire field was in view, and were in position to resume their relation to the line of battle, whenever the scattered fragments of the army could be assembled and formed for an organized resistance to the enemy. In the meantime, it had been decided to mass all the cavalry on the left of the line, opposite to where it had been in the morning. The order came from General Merritt to continue the march in that direction, and the long column led by Lowell turned its head toward the left of the Sixth Corps[40] and formed on the other side of the pike.

Moving across, parallel with the line which had been taken up by that corps, the cavalry was exposed to a galling fire of artillery. One shell took an entire set of fours out of the Sixth Michigan. Not a man left the ranks. The next set closed up the gap. Custer was already there, having been transferred from right to left while the two brigades of the First Division were out on the picket line. Crossing the pike, we passed in front of his division. It was formed in column of regiments, mounted. It is needless to say that they were faced toward the enemy.

Custer himself was riding along the front of his command, chafing like a caged lion, eager for the fray. Devin, with Taylor's battery, had been there for some time, and under the personal direction of General Merritt had been most gallantly resisting the advance of the victorious enemy. The Michigan Brigade took position in front of Custer, Martin's battery next the pike. Lowell with the Reserve Brigade was stationed still farther in advance toward Middletown. The Sixth Corps made its final stand on the prolongation of the cavalry alignment, and from that moment the attacks of the enemy were feeble and ineffective, the battle resolving itself for the time being into an artillery duel, in which Martin's battery took a prominent part.

It could not have been much later than nine o'clock when the two brigades of cavalry arrived. Their coming was opportune. Who can say how much it had to do in stopping the further progress of Early's attack? It is now known that Early dreaded a flanking movement by the body of horse which he saw massing in front of his right flank. The gallant Lowell, who so bravely did his duty and who exhibited in every stage of the battle the highest qualities of leadership, a few hours after his arrival on the left laid down his life for the cause he so valiantly served. He was killed by a bullet from the gun of a sharpshooter in Middletown. He did not live to make a report, and the story never has been told officially of how he marched from right to left at Cedar Creek.

Sheridan had not yet come up, but after his arrival, which he states in his memoirs was not later than ten o'clock, Custer was moved to the right flank, arriving in time to thwart a threatened flanking movement by Gordon and Kershaw. It is evident that every strategic attempt of the enemy, save the morning surprise, was checkmated by the Union cavalry, and it must be remembered that it was the absence of cavalry on the left which rendered the morning surprise possible.

The First Division was now all together with General Merritt personally in command. A part of Lowell's brigade, dismounted, was posted well to the front, the Michigan Brigade, mounted, in its rear. While in this position, having occasion to ride up into the battery to speak to Captain Martin, a sharpshooter in Middletown took a shot at us. The bullet narrowly missed the captain and buried itself in my horse's shoulder. Unlike the shell at Winchester, this wound disabled the old fellow, so that he had to go to the rear and give way to a temporary remount - furnished by the commanding officer of the First Michigan - much to the regret of the old hero, for he was a horse who loved the excitement of battle and relished its dangers.

Thus, for perhaps an hour (it may have been more), we stood in line inviting attack. But the enemy, strongly posted behind fences and piles of logs with two ravines and fences separating us, seemed anxious to let well enough alone. Then Merritt rearranged his line. Devin's brigade was posted next the pike, Lowell in the center, the Michigan Brigade on the extreme left. Martin's battery took position in an orchard on a rising point, which commanded the entire front and sloped off to the rear, so that only the muzzles of the pieces were exposed to the enemy's fire. Directly in front was a section of a battery which Martin several times silenced but which had an aggravating way of coming into action again and making it extremely uncomfortable for us. The First, Sixth and Seventh were formed in line of squadron columns, the Fifth a little to the rear as a reserve and support. A strong line of mounted skirmishers held the front. The left was thrown somewhat forward, menacing the Confederate right.

Soon after the formation was complete and probably not far from eleven o'clock, General Merritt with his staff came along inspecting the line, and halting near Martin's battery, he expressed the most hearty approval of the dispositions that had been made. While he was still talking, a round shot from one of the enemy's guns ricocheted and nearly struck his horse. He was very cool, and gave his view of the situation in a few encouraging words. "The enemy," said he, "is almost as much surprised as we are and does not know what to make of his morning's work, and in my opinion, does not intend to press his advantage, but will retreat as soon as a vigorous assault is made upon his line."

These are, I am sure, almost the precise words uttered to me by General Merritt before Sheridan came up. At least, if he was with the army at the time, certainly General Merritt did not know it. They show what was the feeling in that portion of the army which was not surprised, and which did not fail, from the moment when the first shot was fired in the early morning to the last charge at dusk to keep its face to the foe. General Merritt also suggested, though he did not order it, that I send a regiment to feel of the Confederate right flank. He had an impression that it might be turned.

The Seventh Michigan was sent with instructions to pass by the rear to the left, thence to the front, and attempt to get beyond the flank of the enemy, and if successful to attack. After an absence of about an hour, it returned and the commanding officer reported that he found a line of infantry as far as he deemed it prudent to go. The force in front of the cavalry was Wharton's (Breckinridge's) corps, reinforced by one brigade of Kershaw's division. Early's fear of being flanked by the Union cavalry caused him to strengthen and prolong his right. Rosser's cavalry, for some reason, did not put in an appearance after the dash in the morning.

There was a lull. After the lapse of so many years, it would be idle to try to recall the hours, where they went and how they sped. There was no thought of retreat, slight fear of being attacked. All were wondering what would be done, when cheering and a great commotion arose toward the right. "Sheridan has come; Sheridan has come; and there is to be an advance all along the line," sped from right to left, as if an electric battery had sent the message, so quickly did it fly.

Sheridan did not pass to the left of the pike where the cavalry was, but dashed along in front of the infantry for the purpose of letting the army know that he was there, and give it the inspiration of his presence. History puts in his mouth the words "It is all right, boys; we will whip them yet; we will sleep in our old camps tonight." I was not near enough to hear and do not pretend to quote from personal knowledge, but whatever may have been his exact words, the enthusiasm which they aroused was unmistakable. The answer was a shout that sent a thrill across the valley and whose ominous meaning must have filled the hearts of the Confederates with misgivings. This was the first intimation we had that Sheridan was on the ground, though he says in his memoirs that it was then after midday and that he had been up about two hours.

But the Sixth corps needed no encouragement. Nobly had it done its duty during the entire progress of the battle. Sheridan and his staff, therefore, busied themselves reforming and posting the Nineteenth Corps and strengthening the right where Custer was to be given the post of honor in the grand flanking movement about to begin. An ominous silence succeeded. Even the batteries were still. It was the calm that precedes the storm. To those on the left, it seemed that the dispositions were a long time in making. When one has his courage screwed to the sticking point, the more quickly he can plunge in and have it over the better. The suspense was terrible.

The Michigan Brigade had ample time to survey the field in its front. First, the ground descended abruptly into a broad ravine, or depression, through which ran a small creek. Beyond the top of the opposite ascent was a wide plateau of rather level ground, then another ravine and a dry ditch; then a rise and another depression from which the ground sloped up to a belt of timber stretching clear across the front almost to the pike. In the edge of the timber was the enemy's main line of battle, behind piles of rails and logs. Half way down the slope was a strong skirmish line along a rail fence. Behind the fence, on a knoll, was the battery which had annoyed us so much. The brigade was formed with the First Michigan on the right, the Seventh on the left, the Sixth and Fifth in the center, in the order named. Each regiment was in column of battalions, making three lines of two ranks each. Martin's battery was to continue firing until the cavalry came into the line of fire.

At length, the expected order came. The bugles sounded, "Forward." Simultaneously, from the right to the left the movement began. At first, slowly, then faster. It was a glorious sight to see that magnificent line sweeping onward in the charge. Far, far away to the right it was visible. There were no reserves, no plans for retreat, only one grand, absorbing thought - to drive them back and retake the camps. Heavens, what a din! All along the Confederate line, the cannon volleyed and thundered. The Union artillery replied. The roll of musketry became incessant. The cavalry crossed the first ravine and moving over the level plateau, came into a raking fire of artillery and musketry.

Pressing on, they crossed the second ravine and ditch. The slope was reached, and charging up to the rail fence, the first line of hostile infantry fell back. But the cavalry had gone too fast for the infantry. Sheridan says faster than he intended, for his intention was to swing his right wing and drive the enemy across the pike into the arms of the left wing on the east side; the too swift advance of the First Cavalry division frustrated the plan. The brigade next to the pike, exposed to a galling crossfire, wavered and slowly retired. The entire line then gave way and retreated rapidly, but in good order, to the first ravine, where it halted and reformed. In a short time, the charge was again sounded.

This time the fence was reached. The right of the Sixth Michigan was directly in front of the battery, as was also the First Michigan. General Merritt, who was riding by the side of Major Deane, said; "Major, we want those guns." "All right, we will get them," gallantly responded the major, and through and over the fence rode the brave cavalrymen. The First Michigan made a dash for the battery, but it was not ours this time, for seeing that the Sixth Corps had received a temporary check, the cavalry once more fell back to the nearest ravine, and whirling into line without orders was ready instantly for the last supreme effort, which was not long delayed.

The charge was sounded. The infantry responded with a shout. This time the cavalry pressed right on up the slope. The enemy did not stand to meet the determined assault but gave way in disorder. The line pushed into the woods and then it was every regiment for itself. The First, under Major Duggan, charged toward the pike, but Devin, being nearer reached the bridge first. The Seventh, under Lieutenant Colonel Briggs, charging through a field, captured seemingly more prisoners than it had men. The Sixth, under Major Deane, who knew the country well, did not pause until it reached Buckton's Ford on the Shenandoah River, returning late at night with many prisoners and a battle flag for which Private Ulric Crocker of Troop M received one of the medals awarded by act of Congress. The Fifth, under Major Hastings, charged down a road leading to one of the fords of the Shenandoah, Major Philip Mothersill, with one battalion, going so far that he did not rejoin the command till the next day.[41] Thus ended the Battle of Cedar Creek. Darkness alone saved Early's army from capture. As it was, most of his artillery and wagons were taken.

It is needless to tell how Sheridan broke Early's left by an assault with the Nineteenth Corps and Custer's cavalry at the same moment of the last successful charge upon his right. It was a famous victory, though not a bloodless one. Of the gallant men who went into the fight that morning on the Union side, 588 never came out alive. Three thousand five hundred and sixteen were wounded. Early did not lose so many, but his prestige was gone, his army destroyed, and from that moment, for the Confederacy to continue the hopeless struggle was criminal folly. Cedar Creek was the ending of the campaign in the Shenandoah Valley. There was some desultory skirmishing, but no real fighting thereafter.

Among the wounded were Captain Charles Shier, Jr. and Captain Darius G. Maynard, both of the First Michigan Cavalry. Captain Shier died on the 31st of October. He was wounded in the charge on the Confederate battery. Captain Shier was as gallant an officer as any who periled his life on that famous battlefield; and not only a fine soldier but a polished scholar and an accomplished gentleman as well. He was a distinguished son of the state of Michigan and of the noble university which bears its name. In his life and in his death, he honored both.

Massachusetts remembers the name and reveres the memory of Charles Lowell. Mothers recite to their children the circumstances of his heroic death, and in the halls of Harvard, a tablet has been placed in his honor. Charles Shier is a name which ought to be as proudly remembered in Michigan and in Ann Arbor as is that of Charles Lowell in Massachusetts and in Cambridge. But fate, in its irony, has decreed that the nimbus which surrounds the brow of a nation's heroes shall be reserved for the few whom she selects as types, and these more often than otherwise idealized types chosen by chance or by accident. These alone may wear the laurel that catches the eye of ideality and furnishes the theme for the poet's praise. Others must be content to shine in reflected light or to be forgotten. The best way is to follow William Winter's advice and neither crave admiration nor expect gratitude. After all, the best reward that can come to a man is that intimate knowledge of himself which is the sure foundation of self-respect. The adulation of the people is a fugitive dream, as Admiral Dewey knows now, if he did not suspect it before.

In the original manuscript of the foregoing chapter, written in the year 1886, Lowell was represented as marching without orders from right to left with his own brigade and the Michigan Brigade. In the text, the words "without orders" have been omitted. This is not because my own recollection of the events of that day is not the same now as then, but for the reason that I am reluctant to invite controversy by giving as statements of fact things that rest upon the evidence of my own unsupported memory.

After the manuscript had been prepared, it was referred to General Merritt with a request that he point out any errors or inaccuracies that he might note, as it was intended for publication. This request elicited the following reply:

West Point, December 2, 1886
General J.H. Kidd

My Dear General: So much has been written as to the details of the war that I have stopped reading the war papers in the best magazines, even. An officer writes one month what is to him a truthful account of events and the next month that account is contradicted by three or four in print with dozens of others who content themselves with contradicting it in talk. The account you send me of Cedar Creek is not more accurate than the rest.

The morning of the attack, Lowell's brigade had been ordered to make a reconnoissance on the Middle road. This order was given by me the evening before. The picket line of the First Brigade was attacked before the Reserve Brigade moved out, and Lowell was ordered to hold his brigade in hand to help the First Brigade if the attack was pressed.

Soon after, the fighting on the left of our army was heavy, as shown by the artillery fire, and stragglers commenced coming across towards the back road. These were stopped and formed as far as possible by my headquarters escort - the Fifth U.S. Cavalry. About this time, Devin's brigade (my Second) was ordered to the left of our line to cover and hold the valley pike.

About ten o'clock, the remainder of the First Division was moved to the left of the infantry line and disposed so as to connect with the infantry and cover the valley pike. This was soon done, the Second Brigade (Devin's) occupying the right, the Reserve Brigade (Lowell's) the center, and the First Brigade (Kidd's) the left of the division line of battle.

This is the account of the first part of the battle taken from my report written at the time. The movement of Lowell's brigade and your own by agreement, and without orders, was impossible. We had all been posted where we were as part of a line of battle, and any soldier who took a command without orders from one part of a line to another subjected himself to the penalty of being cashiered, as such action might jeopardize the safety of an army.

The principle of marching to the sound of battle when you are distant and detached and without orders that contemplate the contingency is well defined, but for a commander to leave without orders one part of a line of battle because there appears to be heavier fighting at another is all wrong and could not be tolerated.

I should be glad to renew our acquaintance and talk over the war, though as I have intimated I am sick of the fiction written with reference to it.

Truly yours,

W. MERRITT

General Merritt in his letter omits one clause in his quotation from his report written at the time which seems to me to have an important bearing upon this question. The clause is as follows: "The First Brigade was at once ordered to the support of its picket line." Or to quote the passage in its entirety:

Charles Lowell. Upon hearing of his death at Cedar Creek, General George Armstrong Custer wept. General Sheridan stated "I do not think there was a quality which I could have added to Lowell. He was the perfection of a man and a soldier."

"About 4 a.m. on the 19th, an attack was made on the pickets of the First Brigade near Cupp's Ford, which attack, coupled with the firing on the extreme left of the infantry line, alarmed the camps, and everything was got ready for immediate action. The First Brigade was at once ordered to the support of its picket line, while the Reserve Brigade, which had the night before received orders to make a reconnoissance on the Middle Road, was ordered to halt and await further orders. This brigade had advanced in the execution of its reconnoissance to the picket line, and subsequently acted for a short time with the First Brigade in repelling the attack of the enemy, feebly made on that part of the field. Soon after moving from camp, the heavy artillery firing and immense number of infantry stragglers making across the country to the Back Road from our left showed that it was in that direction the heavy force of the enemy was advancing. The Fifth U.S. Cavalry attached to the division headquarters was deployed across the field, and together with the officers and orderlies of the division staff did much toward preventing the infantry going to the rear. About the same time, the Second Brigade (General Devin) was ordered to move to the left of the line, cover and hold the pike, and at the same time deploy men in that part of the field to prevent fugitives going to the rear."

The rule about moving toward the sound of battle is succinctly stated by General Merritt in his letter and does not admit of controversy. But I may in all fairness call attention to the conditions that existed at the time when it was asserted that Colonel Lowell took the responsibility to move his brigade from the picket line to the rear, if not to the left, and order the First Brigade to follow. The division line of battle of which the three brigades had been a part had been broken up. There was no division line of battle. The First Brigade had been ordered to reinforce its picket line. The Reserve Brigade which on the night before received the order to make a reconnoissance in the morning was held to support the First Brigade and had advanced as far as the picket line. Devin's brigade had been ordered to the valley pike to hold it and deploy men to prevent fugitives going to the rear.

May it not then be said with truth that he was "distant and detached" and "without orders that contemplate the contingency?" The enemy that attacked "feebly" had disappeared. There was in sight no picket line either of the enemy's or of our own. There was visible no line of skirmishers or of battle. The "fighting on the left of our army as shown by the artillery fire" was not only heavy, as described by General Merritt, but indicated clearly by the sound that the army was falling back. Lowell's movement was under the circumstances entirely justifiable. That he moved from the picket line to the rear voluntarily, and that he took the responsibility to order the Michigan Brigade to follow, is as certain as that when the moon passes between the earth and the sun it causes an eclipse.

The march from the picket line to the pike was continuous. There was no halting for formations of any kind. It is quite possible however, that the staff officer who conveyed the order from General Merritt found Lowell in motion in the right direction and delivered the order to him to cover the movement of both brigades. I do not remember receiving any order except the one from Lowell until after reaching the pike.

One more point and this subject, which has been given more space perhaps than it ought to, will be left to the reader. General Merritt's report takes up the matter of arranging the division line of battle with the formation at "about ten o'clock," with the Second Brigade on the right, next to the pike, the Reserve Brigade in the center, and the First Brigade on the left. That was some time after the arrival of the two brigades. The first position taken by the First Brigade was next the pike in rear of Lowell and Devin. Martin's battery was posted originally close to the pike, and it was while there that my horse was shot. I still believe that it was not much after nine o'clock when we first formed on the left of Getty's division. The subsequent rearrangement of the line is referred to in the text and was exactly as described in General Merritt's report.

The following table of killed and wounded in the Michigan Cavalry Brigade in the Shenandoah Valley campaign is compiled from the official records in the office of the adjutant general of Michigan:

Michigan Brigades	First	Fifth	Sixth	Seventh	Total
Winchester	16	8	7	8	39
Shepherd-	1	5	1	0	7

stown					
Middletown	1	--	--	--	1
Smithfield	2	4	2	3	11
On Picket	1	--	--	--	1
Cedar Creek	3	5	6	2	16
By Mosby's Men	--	18	--	--	18
Front Royal	--	2	--	2	4
Newtown	--	4	--	--	4
Tom's Brook	--	--	1	1	2
Berryville	--	--	--	1	1
Total	24	46	17	17	104

Recapitulation--Killed and died of wounds, Shenandoah Valley:

First Michigan	24
Fifth Michigan	46
Sixth Michigan	17
Seventh Michigan	17
Total	104

The following table of killed and wounded in the First Cavalry Division in the Battle of Cedar Creek is taken from the official war records:[42]

First Brigade

Officers and men killed	10
Officers and men wounded	43
Officers and men killed and wounded	43

Second Brigade

Officers and men killed	3
Officers and men wounded	16

Officers and men killed and wounded	19
Reserve Brigade	
Officers and men killed	9
Officers and men wounded	27
Officers and men killed and wounded	36

Total killed and wounded, First Brigade	53
Total killed and wounded Second and Reserve Brigades	55

It is thus seen that the First Brigade lost in killed and wounded within two of as many as both the other brigades - almost fifty per cent of the entire losses of the division. Custer's division of two brigades lost 2 killed and 24 wounded. Powell's division of two brigades lost 1 killed, 8 wounded. In other words, while the entire of the Second and Third Divisions - four brigades - lost but 35 killed and wounded, the Michigan Brigade alone lost 53 in this battle. Thirty-four per cent of the entire losses killed and wounded in the cavalry corps were in this one brigade.[43]

These figures give point to the statement of General Merritt in a communication to the adjutant general of the First Cavalry Division, dated November 4, 1864, that the list of killed and wounded in a battle is presumptive evidence of the degree and kind of service performed.[44] General Merritt also gives the Michigan Brigade credit for "overwhelming a battery, and its supports," in other words capturing the battery.

A MYSTERIOUS WITNESS

IN the latter part of the winter of 1864-65, I was detailed as president of a military commission, called to meet in Winchester to try a man charged with being a spy, a guerrilla, a dealer in contraband goods, and a bad and dangerous man. The specifications recited that the accused had been a member of the notorious Harry Gilmor's band of partisans; that he had been caught wearing citizen's clothes inside the Union lines; and that he was in the habit of conveying quinine and other medical supplies into the Confederacy. He was a mild mannered, inoffensive appearing person who had been an employee of the Baltimore and Ohio Railroad Company.

He appeared under guard before the commission at its daily sessions, accompanied by his counsel, a leading attorney of Winchester, whose learning and ability were not less pronounced than was the quality of his whisky, samples of which he at irregular intervals brought in for the solace, if not for the seduction of the court. It was no more like the article commonly called whisky than Mumm's extra dry is like the pink lemonade of circus time. It had an oily appearance, an aromatic flavor, and the lawyer averred that there was not a headache in a barrel of it, though he was the only one who ever had an opportunity to test the truth of the statement and there is no doubt that he knew.

The prisoner exhibited a surprising degree of sang froid considering the grave crimes with which he was charged, the penalty of conviction for any one of which was death. This attitude of the accused puzzled the commission not a little, for he acted like either a very hardened criminal, or a man who was both conscious of innocence and confident of acquittal, and he did not look like a very bad man.

The case was on trial when the army moved. General Sheridan seemed to lay much stress on the matter, for he refused the request of the president of the commission to be relieved in order to rejoin his regiment. A personal letter from General Merritt to General Forsythe, chief-of-staff, making the same request was negative, and an order issued directing the commission to remain in session until that particular case was disposed of, and providing that such members as should then desire it, be relieved and their places filled by others.

During the progress of the trial, the commission was informed that a very important witness had been detained under guard, by order of General Sheridan, in order that his testimony might be taken. On the witness's first appearance, it was noticed that the guard detail was very careful to give him no opportunity to escape. He proved to be a person of most noticeable appearance. Rather above than under six feet, well-built, straight, athletic, with coal-black hair worn rather long, a keen, restless black eye, prominent features, well dressed, and with a confident, devil-may-care bearing, he was altogether a most striking figure.

His name was Lemoss; his testimony to the point and unequivocal. He acknowledged having been a guerrilla, himself. He had, he said, been a member of Gilmor's band and of other equally notorious commands. He had deserted and tendered his services as a scout and they had been accepted by General Sheridan. He swore that he knew the prisoner; had seen him serving with Gilmor; and knew that he had been engaged in the practices charged.

After this witness had given his testimony, the court saw no more of him, but he left a very bad impression on the minds of the members, and there was not one of them who did not feel and give voice to the suspicion that there was something mysterious about him which was not disclosed at the trial. When news of the assassination of the president came to Winchester, all wondered if he did not have something to do with it, and the name "Lemoss" was instantly on the lips of every one of us. He had, in the meantime disappeared.

When I met General Sheridan in Petersburg after the surrender, and he inquired what disposition had been made of that case, I told him of the distrust of the principal witness and that it was the unanimous opinion of the commission that the witness was a much more dangerous man than the prisoner. The general smiled and remarked, rather significantly I thought, that he kept Early's spies at his headquarters all winter, letting them suppose that they were deceiving him, and that before the army moved he had sent them off on false scents.

The inference I drew from the conversation was that Lemoss was one of those spies and that the trial was a blind for the purpose of keeping him where he could do no harm, without letting him know that he was under suspicion. Nothing more was said about the matter, and I presume that, at the time, General Sheridan did not know what had become of Lemoss.

Soon after the Grand Review, my regiment was ordered to the west, and while en route to Leavenworth, Kansas, I stopped overnight in St. Louis. When reading the morning paper at the breakfast table, I came upon an item which was dated in some New England city, Hartford or New Haven, I think, stating that a man by the name of Lemoss, who had been a scout at Sheridan's headquarters in the Shenandoah Valley, had been arrested by the police in the city in question, and papers found on his person tending to show that he had been in some way implicated in the plot to assassinate President Lincoln.

This recalled to my mind the surmises in Winchester on the day of the event, and also the hint thrown out by General Sheridan in reply to my question in Petersburg. I cut the slip out, intending to keep it, but before my return to the states a long time afterwards had both lost it and temporarily forgotten the circumstance. It was not until many years had elapsed and I began to think of putting my recollections of the war into form for preservation that all these things came back to my mind. I have often told the story to comrades at regimental or army reunions. The conjectures of the members of the military commission; the suggestion of General Sheridan that Lemoss was a Confederate spy; and the newspaper clipping in St. Louis; all seemed so coincident as to form a pretty conclusive chain of evidence connecting the Winchester witness with the conspiracy. I never learned what was done with him after the arrest in New England.

The Grand Review. At 9:00 a.m. on May 23, 1865, a signal gun fired a shot and General George Meade led 80,000 men of Army of the Potomac down the streets of Washington from Capitol Hill, down Pennsylvania Avenue and past crowds numbering in the thousands. The infantry marched with twelve men across the road, followed by the artillery, then an array of cavalry regiments that stretched for another seven miles.

Recently, when consulting Sheridan's memoirs to verify my

own remembrance of the dates of certain events in the Shenandoah campaign, what was my surprise to find that the purport of a passage bearing directly upon this subject had entirely escaped my attention on the occasion of a first reading soon after the book appeared. On page 108, volume 2, appears the following:

"A man named Lomas, who claimed to be a Marylander, offered me his services as a spy, and coming highly recommended from Mr. Stanton, who had made use of him in that capacity, I employed him. He made many pretensions, was more than ordinarily intelligent, but my confidence in him was by no means unlimited. I often found what he reported corroborated by Young's men, but generally, there were discrepancies in his tales which led me to suspect that he was employed by the enemy as well as by me. I felt however, that with good watching, he could do me very little harm, and if my suspicions were incorrect, he might be very useful, so I held on to him."

"Early in February, Lomas was very solicitous for me to employ a man who he said had been with Mosby, but on account of some quarrel had abandoned that leader. Thinking that with two of them I might destroy the railroad bridge east of Lynchburg, I concluded after the Mosby man had been brought to my headquarters, by Lomas about 12 o'clock one night, to give him employment at the same time informing Colonel Young that I suspected their fidelity and that he must test it by shadowing their every movement. When Lomas's companion entered my room, he was completely disguised, but on discarding the various contrivances by which his identity was concealed he proved to be a rather slender, dark-complexioned, handsome young man, of easy address and captivating manners. He gave his name as "Renfrew," answered all questions satisfactorily, and went into details about Mosby and his men which showed an intimacy with them at some time. I explained the work I had laid out for them, * * * * * They assented and it was arranged that they should start the following night."

"Meantime, Young had selected his men to shadow them, and two days later they reported my spies as being concealed in Strasburg without making the slightest effort to continue on their mission. On the 16th of February, they returned and reported their failure, telling so many lies as to remove all doubt as to their double-dealing. Unquestionably, they were spies, but it struck me that through them I might deceive Early as to the time of opening the spring campaign. I therefore retained the men without even a suggestion of my knowledge of their true character. Young, meantime, kept close watch over all their doings."

General Sheridan then, after giving a summary of the scattered locations of the various portions of Early's army continues as follows:

"It was my aim to get well on the road before Early could collect these scattered forces, and as the officers had been in the habit of amusing themselves during the winter by fox hunting, I decided to use the hunt as an expedient for stealing a march on the enemy, and had it given out that a grand fox chase would take place on the 29th of February. Knowing that Lomas and Renfrew would spread the announcement south, they were permitted to see several red foxes as well as a pack of hounds which had been secured for the spurt and were then started on a second expedition to burn the bridges. Of course they were shadowed, and two days later were arrested in Newtown. On the way north, they escaped from their guards when passing through

Baltimore, and I never heard of them again, though I learned that after the assassination of Mr. Lincoln, Secretary Stanton strongly suspected his friend Lomas of being associated with the conspirators, and it then occurred to me that the good looking Renfrew may have been Wilkes Booth, for he certainly bore a strong resemblance to Booth's pictures."

There is no doubt that "Lemoss," the witness, and the "Lomas" of General Sheridan's narrative, were one and the same person. When he wrote the account from which the foregoing is an extract, General Sheridan had probably forgotten about leaving the spies in Winchester under guard where they remained until he was well on his way towards Appomattox. After giving his testimony, Lomas and Renfrew were sent north under guard by General Hancock, Sheridan's successor as commander of the Middle Military Division, and making their escape as explained in Sheridan's narrative, Wilkes Booth, alias Renfrew, was able to carry out his part of the plot.

It is also quite probable that Lomas's part in the conspiracy was to assassinate either General Sheridan or Secretary Stanton, but that the scheme was interrupted by the detention of the two spies in Winchester coupled with the unexpected opening of the spring campaign. It is likely that the arrest of the two conspirators led to a postponement of the date of the assassination and that the scope of the plot as originally conceived in the fertile brain of Booth was very much abridged.

There was never in my own mind a particle of doubt, from the moment we heard the news of the president's death, that the man Lomas or Lemoss had something to do with it. The fact that he was on terms of intimacy with Secretary Stanton and contrived to be stationed at Sheridan's headquarters seems to point conclusively to the part he was to play in the tragedy. At that time, Sheridan was considered, perhaps, the most dangerous enemy the Confederacy had to fear, and his name must have been high up in the list of those marked by the conspirators for assassination.

John Wilkes Booth. When family friend John T. Ford opened 1,500-seat Ford's Theatre on November 9 in Washington, D.C., Actor John Wilkes Booth was one of the first leading men to appear there, playing in Charles Selby's *The Marble Heart*. Booth portrayed a Greek sculptor, making marble statues come to life. Abraham Lincoln was present, watching the performance of his eventual assassin.

An amusing incident occurred as this trial neared its close. The defense asked to have William Prescott Smith, master of transportation of the Baltimore and Ohio Railroad, summoned as a witness. His residence was Baltimore and he was summoned by wire, the telegram bearing the name of General Hancock, commander of the department. Mr. Smith did not want to come to Winchester and urged the commission to go to Baltimore. Failing to secure acquiescence in that proposition, he suggested as a compromise that the commission meet him halfway by going to Harper's Ferry. This was agreed to and on the appointed day, the commission took passage on a special train consisting of a locomotive and one passenger coach taking along the prisoner and a guard.

Harper's Ferry was reached a little after dark, and a messenger from Mr. Smith met us with the compliments of that gentleman and a request that we proceed to his private car. The invitation was accepted and the party was received by the railroad magnate with every manifestation of welcome and a courtesy that seemed to be entirely unaffected. It was found that the most generous and thoughtful provision had been made for our comfort. The colored chef prepared a dinner which would have tickled the palate of an epicure, much more those of a quartet of hungry officers directly from the front. There were champagne and cigars in abundance of a quality such as would have been good enough had General Hancock himself been the guest. The host was courtesy itself, an excellent raconteur, a good fellow, and a gentleman. He could not have treated the president and his cabinet with more distinguished consideration that that with which he honored that little party of volunteer officers.

Late in the evening, his testimony was taken and he gave the prisoner a very good character. We slept in his car, and in the morning had a breakfast that suitably supplemented the elegant dinner. Some more choice cigars, and then Mr. Smith's private car was attached to an ingoing train and he departed for Baltimore. At the very last moment before his train started, Mr. Smith said:

"Pardon me, gentlemen, but it is too good a joke to keep and I am sure that you will appreciate it now better than you would have done last night. When you wired me to come, you know, General Hancock's name was signed to the telegram. I supposed I was to entertain him and don't mind telling you, frankly, that the dinner was provided with especial reference to his supposed partiality for the good things of life. I don't mean to say I would not have done the same thing for you. I certainly would now that I know you, but all the same, please say to the general that I expected him and regret much that he was not one of the party so that I might have had the pleasure of entertaining him as well as yourselves. And by the way, he continued, when I urged you to come to Baltimore, it had been arranged that the mayor and a large number of prominent citizens of the city were to meet you at a banquet to have been given at the Eutaw House in honor of General Hancock."

The refined courtesy of the gentleman was something that has been rarely surpassed. Mr. Smith was a thoroughbred.

A MEETING WITH MOSBY

AT the time of the surrender of Lee and the fall of Richmond, about the only Confederate force in the Shenandoah Valley was Mosby's band. The last of Early's army had been swept away by Sheridan's advance, led by Custer, and for the first time since 1860, that beautiful valley was free from the movements of armed forces confronting each other in hostile array. The bold and dashing partisan was, however, capable of doing much mischief and it was thought best by General Hancock to treat with him and see if he would not consent to a cessation of hostilities, and possibly take the parole.

Accordingly, an agreement was made to meet him at Millwood, a little town a few miles distant from Winchester and near the mountains. General Chapman, a cavalry officer, was selected to conduct the negotiations and with an escort of two regiments left early on the morning of the day designated for the rendezvous agreed upon. Not yet having been relieved from duty there, I readily obtained permission to accompany the expedition. I was early in the saddle and joining a party of staff officers, struck across country, arriving at about the same time as the escort which took the main road.

The region to which we were going was one of the favorite haunts of Mosby and his men, and it produced a queer sensation to thus ride peacefully through a country where for four long years, the life or liberty of the Union soldier caught outside the lines had been worth not a rush, unless backed by force enough to hold its own against an enemy. There never had been a time since our advent into this land of the philistines (a land literally flowing with milk and honey) when we could go to Millwood without a fight, and here we were going without molestation, right into the lair of the most redoubtable of all the partisan leaders.

John Singleton Mosby. Of his exploits in the war, Mosby wrote. "It is a classical maxim that it is sweet and becoming to die for one's country; but whoever has seen the horrors of a battlefield feels that it is far sweeter to live for it."

But Mosby's word was law in that section. His fiat had gone forth that there was to be a truce, and no Union men were to be molested until it should be declared off. There was, therefore, no one to molest or make us afraid. No picket challenged. Not a scout or vidette was seen. The country might have been deserted, for all the indications of life that could be heard or seen. The environment seemed funereal, and the ride could hardly be described as a cheerful one. Each one was busy with his own thoughts. All wondered if the end had really come, or was it yet afar off? Lee had surrendered but Johnson had not. Would he?

The chief interest, for the time being, however, centered in the coming interview with Mosby under a flag of truce. If he could be prevailed upon to take the parole, there would not be an armed Confederate in that part of Virginia. It had been expected that he would be there first, but he was not and his arrival was eagerly awaited. The escort was massed near a large farm house, the owner of which was very hospitable and had arranged to give the two commands a dinner.

The officers were soon dispersed in easy attitudes about the porches and lawn or under the shade of friendly trees, smoking and chatting about the interesting situation. Eager glances were cast in the direction from which our old foe was expected to come, and there was some anxiety lest he should fail to meet the appointment after all. But at length, when the forenoon was pretty well spent, the sound of a bugle was heard. All sprang to their feet. In a moment, the head of a column of mounted men emerged from a woody screen on the high ground toward the east, as though coming straight out of the mountain, and presently, the whole body of gray troopers came into view.

It was a gallant sight, a thrilling scene, for all the world like a picture from one of Walter Scott's novels; and to the imagination, seemed a vision of William Wallace or of Rob Roy. The place itself was a picturesque one - a little valley nestling beneath the foothills at the base of the mountains whose tops towered to the sky. Hills and wooded terraces surrounded it, shutting it in on all sides, obstructing the view and leaving the details of the adjacent landscape to the imagination.

Mosby evidently had arranged his arrival with a view to theatric effect - though it was no mimic stage on which he was acting - for it was to the sound of the bugle's note that he burst into view, and like a highland chief coming to a lowland council, rode proudly at the head of his men. Finely uniformed and mounted on a thoroughbred sorrel mare whose feet spurned the ground, he pranced into our presence. Next came about sixty of his men, including most of the officers, all, like himself, dressed in their best and superbly mounted. It was a godly sight to see.

General Chapman advanced to meet the commander as he dismounted, and the two officers shook hands cordially. There were then introductions all around and in a few moments, the blue and the gray were intermingling on the most friendly terms.

It was difficult to believe that we were in the presence of the most daring and audacious partisan leader, at the same time that he was one of the most intrepid and successful cavalry officers in the Confederate service. He was wary, untiring, vigilant, bold, and no Federal trooper ever went on picket without the feeling that this man might be close at hand watching to take advantage of any moment of unwariness. He had been known, in broad daylight, to dash right into Federal camps, where he was outnumbered a hundred to one, and then make his escape through the fleetness of his horses and his knowledge of the byroads. On more than one occasion, he had charged through a Union column, disappearing on one flank as quickly as he had appeared on the other. His men, in Union garb, were often in our camps mingling unsuspected with our men or riding by their side when on the march.

We were prepared to see a large, fierce looking dragoon, but instead beheld a small, mild-mannered man not at all like the ideal. But, though small, he was wiry, active, restless and full of fire. "How much do you weigh, colonel?" I asked as I shook his hand and looked inquiringly at his rather slender figure. "One hundred and twenty-eight pounds," said he. "Well, judging from your fighting reputation, I looked for a two hundred pounder, at least," I replied.

His spare form was set off by a prominent nose, a keen eye and a sandy beard. There was nothing ferocious in his appearance, but when in the saddle he was not a man whom one would care to meet single-handed. There was that about him which gave evidence of alertness and courage of the highest order.

It was astonishing to see officers of Mosby's command walk

up to Union officers, salute and accost them by name. "Where did I meet you?" would be the reply. "There was no introduction. I met you in your camp, though you were not aware of it at the time."

Major Richards, a swarthy-looking soldier, remarked to me that he was once a prisoner of the Fifth and Sixth Michigan Cavalry. He was captured near Aldie in the spring of 1863, and made his escape when the Michigan regiments were on the march back to Fairfax Court House in the night, when his guards were not noticing, by falling out of the column and boldly ordering his captors to "close up" as they were coming out of a narrow place in the road when the column of fours had to break by twos. In the darkness and confusion, he was mistaken for one of our own officers. After he had seen the column all "closed up," he rode the other way.

After awhile, the farmer called us in to dinner, and the blue and the gray were arranged around the table in alternate seats. I sat between two members of the celebrated Smith family. One of them, R. Chilton Smith, was a relative of General Lee, or of his chief-of-staff, a young man of very refined manners, highly educated and well bred. He sent a package and a message by me to a friend in Winchester, a commission that was faithfully executed.

The other was the son of Governor, better known as "Extra Billy" Smith, of Virginia; a short, sturdy youth, full of life and animation and venom. "Mosby would be a blanked fool to take the parole," said he, spitefully. "I will not, if he does." "But Lee has surrendered. The jig is up. Why try to prolong the war and cause further useless bloodshed?" "I will never give up so long as there is a man in arms against your Yankee government," he replied. "But what can you do? Richmond is ours." "I will go and join Joe Johnston." "It is a question of but a few days, at most, when Sherman will bag him." "Then I will go west of the Mississippi, where Kirby Smith still holds the fort." "Grant, Sherman, Sheridan and Thomas will make short work of Kirby

Smith." "Then, if worst comes to worst," he hotly retorted, "I will go to Mexico and join Maximilian. I will never submit to Yankee rule; never."

I greatly enjoyed the young man's fervor and loyalty to his cause, and in spite of his bitterness, we took quite a liking to each other and on parting, he was profuse in his expressions of regard and urged me cordially not to forget him should fortune take me his way again.

A day or two later, I was ordered to Petersburg, and soon thereafter was in Richmond, Johnston having, in the meantime, surrendered. In the evening of the day of my arrival, after having visited the points of interest, Libby prison, the burnt district, the state house, etc., I was in the office of the Spotswood Hotel where were numbers of Federal and Confederate soldiers chatting pleasantly together, when I was saluted with a hearty "Hello; how are you, colonel!" and, on looking around, was surprised as well as pleased to see my young friend of the Millwood Conference.

The Spotswood Hotel was a luxurious hotel in Richmond, Virginia. It survived the Civil War, but was destroyed in a fire on Christmas, 1870.

I was mighty glad to meet him again and told him so, while he seemed to reciprocate the feeling. There was a cordial shaking of hands, and after the first friendly greetings had been exchanged I said; "But what does this mean? How about Mexico and Maximilian? Where is Mosby? What has been going on in the valley? Tell me all about it." "Mexico be blanked" said he. "Mosby has taken the parole and so have I. The war is over and I am glad of it. I own up. I am subjugated." The next day I met him again. "I would be only too glad to invite you to our home and show you a little hospitality," said he, "but your military governor has taken possession of our house, father has run away, and mother is around among the neighbors."

I assured him of my appreciation of both his good will and of the situation, and begged him to be at ease on my account. He very politely accompanied me in a walk around the city and did all he could to make my stay agreeable.

I never saw him afterwards. When in Yorktown in 1881, I made inquiry of General Fitzhugh Lee about young Smith and learned that he was dead. I hope that he rests in peace, for although a rebel and a guerrilla, as we called them in those days, he was a whole hearted, generous and courageous foe, who though but a boy in years was ready to fight for the cause he believed in, and in true chivalrous spirit, grasp the hand of his former adversary in genuine kindness and good-fellowship.

One other incident of the Millwood interview is perhaps worth narrating. A bright eyed young scamp of Mosby's command mounted the sorrel mare ridden by his chief, and flourishing a roll of bills which they had probably confiscated on some raid into Yankee territory, rode back and forth in front of the lawn, crying out; "Here are two hundred dollars in greenbacks which say that this little, lean, sorrel mare of Colonel Mosby's, can outrun any horse in the Yankee cavalry." The bet was not taken.

THE END

ROLL OF HONOR - LIST OF KILLED IN ACTION

Following is a list of those killed in action, or who died of wounds received in action in the four regiments which constituted the Michigan Cavalry Brigade, commanded by General George Armstrong Custer, in the Civil War of 1861-65. It constitutes a veritable roll of honor:

FIRST MICHIGAN CAVALRY

NAME AND POSITION	COMPANY	BATTLE	DATE OF DEATH
Adams, William, Private	H	Cold Harbor	June 1, 1864
Alcott, Richard, Private	L	Cedar Mountain	August 9, 1862
Altenburg, William, Corporal	B	Bull Run	August 30, 1862
Andrus, John, Private	K	Winchester	September 19 1864
Anson, Elisha B., Sergeant	E	Haw's Shop	May 28 1864
Babcock, Edwin H., Private	K	Gettysburg	July 3 1863
Bachman, Robert, Sergeant	G	Appomattox	April 9 1865
Banker, Edward S., Private	C	Trevilian Station	June 11 1864

Barney, Lorenzo J., Private	A	Yellow Tavern	May 11 1864
Bartlett, Orrin M., Lieutenant	H	Five Forks	April 1 1865
Bateman, Cyrus A., Corporal	M	Shenandoah Valley	August 111864
Battison, William, Sergeant	H	Piedmont	April 17 1862
Bell, Charles S., Private	E	Todd's Tavern	April 7 1864
Beloir, Michael, Sergeant	B	Trevilian Station	June 11 1864
Bentley, Augustus W., Corporal	I	Gettysburg	July 3 1864
Brown, Dexter, Corporal	E	Yellow Tavern	June 11 1864
Blount, Lemuel K., Private	A	Yellow Tavern	June 11 1864
Bovee, John S., Sergeant	F	Gettysburg	July 3 1864
Brevoort, William M., Captain	K	Cold Harbor	June 1 1864
Brewer, Charles E., Private	A	Bull Run	August 301862
Brodhead, Thornton F., Colonel		Bull Run	August 301862
Bucklin, Lyman D., Private	C	Unknown	May 13 1863
Buhl, Augustus F., Captain	C	Shepherdstown	August 251864
Butler, Abner K., Private	F	Middletown	April 4 1862
Byscheck, John, Private	C	Dinwiddie Courthouse	March 30 1865
Campeau, Eli, Private	K	Unknown	Died July 3 1865
Carr, Alpheus W., Captain	I	Trevilian Station	June 12 1864
Chatfield, William H., Private	B	Bull Run	August 301862
Chilson, Alphonso	I	Yellow Tavern	May 11

W., Sergeant			1864
Chittenden, Adelbert, Private	G	Gettysburg	July 3 1863
Cicotte, David, jr., Private	C	Winchester	February 23 1865
Clarke, John R., Private	K	Winchester	September 19 1864
Cole, Benjamin, Chief Bugler		Winchester	September 19 1864
Colles, David W., Private	I	Unknown	May 26 1865
Crawford, Charles C., Private	M	Todd's Tavern	May 7 1864
Crosby, Henry, Private	E	Unknown	Died June 1 1864
Cummings, George W., Private	A	Yellow Tavern	May 11 1864
Cunningham, Barnabas, Private	A	Smithfield	August 29 1864
Davis, Joseph, Private	I	Unknown	Died June 20 1864
Davison, Joseph, Private	G	Unknown	Died April 7 1865
Dibble, Darius, Private	L	Cedar Mountain	August 9 1862
Dorsay, John, Private	B	Appomattox	April 9 1865
Durkee, Robert, Private	K	Bull Run	August 30 1862
Eagle, Ellwood, Private	H	Cold Harbor	June 1 1864
Eastman, Oscar A, Sergeant	I	Winchester	September 19 1864
Eaton, William O, Private	H	Accident	October 28 1862
Edgerton, George W, Private	L	Beaver Dam	May 9 1864
Elliott, William R, Captain	C	Fairfield Gap	July 4 1863
Ellis, Henry, Private	L	Cedar Mountain	August 9 1862

Ensign, Leroy, Private	M	Winchester	May 4 1862
Fisher, Peter, Private	E	Trevilian Station	June 11 1864
Follett, Irving B, Sergeant	L	Winchester	September 19 1864
Foss, Andrew, Private	I	On Picket	December 14 1864
Frost, Joel, Corporal	L	Cedar Mountain	August 9 1862
Falcher, John, Private	K	Gettysburg	July 3 1863
Gillett, George M, Corporal	B	Hagarstown	July1863
Gordon, Alexander, Corporal	H	Winchester	September 19 1864
Graves, Benjamin F, Private	A	Cold Harbor	June 1 1864
Grimes, Micah, Private	M	Unknown	September 2,1864
Handy, Lucius F, Private	F	Todd's Tavern	May 7 1864
Hart, Lorenzo, Corporal	L	Dinwiddie Courthouse	March 30 1865
Hicks, Charles Eugene, Private	F	Fort Scott	January 12 1863
Hobbs, David, Private	B	Bull Run	August 301862
Hoffman, Peter, Corporal	B	Trevilian Station	June 12 1864
Hough, Albert or Robert, Private	H	Unknown	Died April 8 1865
Hovey, Henry, Private	A	Unknown	Died June 18 1864
Hughes, Patrick H., Corporal	E	Trevilian Station	June 11 1864
Hutton, Thomas, Private	C	Snicker's Ferry	March 26 1862
Hymen, Ralph, Private	I	Bull Run	August 301862
Iott, Harrison, Private	I	Harper's Ferry	August

			311864
Irwin, H. II., Private	F	Unknown	September 5 1864
Irwin, Stephen, H., Sergeant	I	Old Church	May 30 1864
Jackson, Albert T., Captain	F	Winchester	September 19 1864
Jackson, William, Private	K	Rapidan River	September 14 1863
Jacob, Henry, Private	A	Yellow Tavern	May 11 1864
Jacobs, George A., Private	I	Yellow Tavern	May 11 1864
Jayne, William H., Sergeant	G	Unknown	Septembe r23 1863
Kidder, Hiram O., Private	A	Haw's Shop	May 28 1864
Keferly (Keferle) Frank, Private	H	Bull Run	August 301862
Kilbride (Kilride) William, Private	H	Piedmont	April 17 1862
Kling, Henry, Private	G	Gettysburg	July 3 1863
Kroop, Albert, Private	H	Unknown	April 8 1865
Lambert, Jacob, Private	K	Unknown	June 16 1864
Lewis, Lewis J., (Lucius) Private	K	Unknown	June 15 1864
Long (Lozo) Henry, Private	I	Cedar Creek	October 19 1864
Longdo, Jeremiah, Private	B	Unknown	June1864
Lyon, James B., Sergeant	L	Unknown	June 4 1864
McDermott, James, Corporal	A	Bull Run	August 291862
McElheny, James S., Captain	G	Fairfield Gap	July 4 1863
Manuel, Peter, Private	K	Unknown	Died July 29 1864

Marshner, Frank A., Private	A	Winchester	September 19 1864
Martin, David, Private	C	Fairfield Gap	July 4 1863
Mathews, Samuel M., Private	A	Indians	August 13 1865
Merriam, John G., Private	K	Bull Run	August 30 1862
Michaels, William H., Private	C	Trevilian Station	June 12 1864
Miller, John, Private	E	Yellow Tavern	May 11 1864
Moran, Thomas, Private	A	Haw's Shop	May 28 1864
Morse, Sidney G., First Sergeant	M	Bull Run	August 30 1862
Moss, Daniel B., Sergeant	A	Dinwiddie Courthouse	March 30 1865
Murray, Elias M., Private	M	Yellow Tavern	May 11 1864
Nesbit, James, Private	L	Trevilian Station	June 12 1864
Orth, Adam, Private	A	Haw's Shop	May 28 1864
Owen, Perry, Private	F	Cedar Creek	October 19 1864
Phelps, Ralph Z., BattalionAdjutant		Accident	April 1 1862
Pierce, Henry C., Sergeant	B	Bull Run	August 30 1862
Piper, Leo, Sergeant	C	Five Forks	April 1 1865
Pixley, John, Private	K	Gettysburg	July 3 1863
Power (Tower) Mortimer F.,Private	C	Trevilian Station	June 12 1864
Pulver, Andrew J., Lieutenant	A	Trevilian Station	June 12 1864
Price, William H., Private	L	Trevilian Station	June 11 1864
Ransom, William W.,	K	Unknown	Died August 3

Sergeant			1864
Reed, Charles D., Private	K	Smithfield	August 291864
Rennan, Frederick, Private	E	Cold Harbor	June 1 1864
Reynolds, Samuel W., Private	F	Yellow Tavern	May 11 1864
Rhoades, Willard, Q. M. Sergeant	B	Centerville	November 6 1863
Robertson, William, Sergeant	I	Wilderness	May 6 1864
Robins, Charles H., Private	A	Winchester	September 19 1864
Robinson, George W., Lieutenant	A	Winchester	September 19 1864
Rose, William L., Com. Sergeant	G	Unknown	December 25 1864
Rush, Thomas, Private	K	Unknown	Died July 13 1864
Ryder, Alfred G., Corporal	H	Gettysburg	July 3 1863
Saulsbury, Charles, Private	K	Bull Run	August 301862
Sawyer, Henry O., Private	I	Gettysburg	July 3 1863
Schintzler, Leonard, Private	H	Old Church	May 30 1864
Shanahan, Thomas, Corporal	H	Fountaindale	July 4 1863
Shaughnessy, William, Private	B	Bull Run	August 301862
Shier, Charles, jr., Captain	K	Cedar Creek	October 19 1864
Sitts, Charles, Private	L	Fairfield Gap	July 4 1863
Smith, Marcus, Private	I	Accident	May 20 1864
Snyder, Charles F., Captain	F	Hagarstown	July 6 1863
Stanley, Henry C., Private	F	Five Forks	April 1 1865

Stewart, William, Private	C	Winchester	September 19 1864
Sterling, Richard, Hospital Steward		Unknown	November 6, 1864
Teebles, William H.,Private	C	Brentsville	June 7 1863
Thomas, Abel, Private	H	Trevilian Station	June 11 1864
Thomas, Benjamin, Private	B	Bull Run	August 301862
Thomas, Cassius M., Private	M	Wilderness	May 6 1864
Thomas, Samuel H., Private	C	Gettysburg	July 3 1863
Truesdale, Lewis B., Sergeant	K	Winchester	September 19 1864
Vance, George, Private	K	Bull Run	August 301862
Vandecar, Thomas H., Private	L	Unknown	Died May 26 1865
Vashaw, John, Private	K	Bull Run	August 301862
Warren, Robert S., Lieutenant	C	Trevilian Station	June 12 1864
Waterman, W. E., First Sergeant	H	Unknown	Died June 20 1864
Watson, Colbert R., Sergeant	L	Falling Waters	July 14 1863
Welch, Jay Michael, Private	A	Winchester	August 111864
Welton, Ransom W., Private	E	Gettysburg	July 3 1863
Wescott, James M., Private	K	Haw's Shop	May 28 1864
Whitney, Ambrose, Private	H	By Accident	March 7 1862
Whitney, George C., Lieutenant	F	Five Forks	April 1 1865
Wideroder, John C., Private	F	Trevilian Station	June 12 1864
Wilcox, Alonzo	H	Brentsville	June 7

W., Sergeant			1863
Wilcox, Philip, jr., Private	L	Gettysburg	July 3 1863
Williams, Isaac, Private	K	Yellow Tavern	May 11 1864
Winfield, George D., Corporal	D	Salem	April 1 1862
Warwick, William, Private	K	Yellow Tavern	May 11 1864
Wieg, Orren, Private	L	Falling Waters	July 14 1863

FIFTH MICHIGAN CAVALRY

NAME AND POSITION	COMPANY	BATTLE	DATE OF DEATH
Ackerman, Hiram, Corporal	A	Haw's Shop	May 28 1864
Allen, Nelson A., Private	D	Gettysburg	July 3 1863
Allison, George S., Private	B	Winchester	September 19 1864
Alverson, Thomas J., Private	G	Winchester	September 19 1864
Anderson, Alfred C., Private	D	Boonesborough	July 8 1863
Atherholt, Peter, Private	F	Winchester	September 19 1864
Axtell, Benjamin F., Captain	F	Yellow Tavern	May 11 1864
Ball, William, Private	M	Haw's Shop	May 28 1864
Barbour, Frank A., Sergeant	A	Gettysburg	July 3 1863
Barse, Horace S., Corporal	E	Gettysburg	July 3 1863
Beebe, Henry C., Corporal	A	Morton's Ford	November 27 1863
Bemis, Andrew J., Private	K	Haw's Shop	May 28 1864

Benning, John, Private	F	Unknown	Died August 7 1865
Bishop, Abraham, Private	B	Haw's Shop	May 28 1864
Bliss, Henry G., Private	I	Raccoon Ford	September 16 1863
Brennan, William, Sergeant	B	Monterey Gap	July 4 1863
Brink, Simeon L., Private	B	Buckland Mills	October 19 1863
Brown, Clifton E., Private	A	Haw's Shop	May 28 1864
Brown, William, Private	H	Trevilian Station	June 11 1864
Buell, John, Private	K	Gettysburg	July 3 1864
Burdick, Reuben, Private	I	Haw's Shop	May 28 1864
Burnett, Henry, Corporal	D	Shepherdstown	August 291864
Burson, Joseph, Private	L	Haw's Shop	May 28 1864
Busley, Levi, Private	M	Richmond	March 1 1864
Cathcart, Albert J., Private	B	Unknown	Died July 5 1864
Chapman, Edward, Private	E	Buckland Mills	October 19 1863
Chart (Chant), Private	G	Salem	October 23 1864
Clark, Frederick, Private	F	Trevilian Station	June 11 1864
Clark, Henry, Private	M	Dinwiddie Courthouse	April 4 1865
Clyde, Charles B., Private	M	Berryville by Guerrillas	August 191864
Colf, Levinas, Private	K	Wilderness	May 6 1864
Comte, Victor E., Wagoner	C	Unknown	Died July 11 1864
Connor, James,	A	Morton's Ford	November 27

Name	Company	Place	Date
Private			1863
Corcelins, Frederick, Private	K	Gettysburg	July 3 1863
Corser, Augustus F., Private	C	Stevensburg	October 30 1863
Coston, Peter, Private	M	Berryville by Guerrillas	August 191864
Craft, Charles, Private	M	Berryville by Guerrillas	August 191864
Day, Alpheus G., Corporal	E	Berryville by Guerrillas	August 191864
Dean, Henry J., Private	D	Yellow Tavern	May 12 1864
Decker, George R., Private	K	Trevilian Station	June 11 1864
Dell, Martin V., Private	H	Trevilian Station	June 11 1864
Derwin, Lewis, Private	C	Winchester	September 19 1864
Dockham, Reuben K., Sergeant	C	Unknown	Died June 18 1864
Duffey, James, Corporal	F	Falmouth	August 4 1863
Eggleston, Andrew J., Sergeant	K	Unknown	Died July 1 1864
Essler, Samuel K., Private	C	Berryville by Guerrillas	August 191864
Evans, Andrew R., Private	A	Gettysburg	July 3 1863
Felt, John, Private	H	Trevilian Station	June 11 1864
Ferry, Noah H., Major		Gettysburg	July 3 1863
Fox, Josiah, Sergeant	M	Trevilian Station	June 11 1864
Friday, Adolph, Private	F	Yellow Tavern	May 11 1864
Gale, Henry D., Corporal	C	Trevilian Station	June 11 1864
Garvelink,	I	Haw's Shop	May 28

Herman, Corporal			1864
Gibbs, Levi, Private	G	Gettysburg	July 3 1863
Gillett, William H., Corporal	K	Wilderness	May 6 1864
Granger, Edward G., Lieutenant	C	Front Royal	August 16 1864
Gudith, John D., Corporal	D	Cedar Creek	October 19 1864
Hammond, Smith Sergeant	G	Brandy Station	October 11 1863
Hanly, Richard, Private	E	Middletown	August 15 1864
Harmon, Allen M., Lieutenant	B	By Accident	April 20 1863
Henry, Alfred A., Private	C	Berryville by Guerrillas	August 19 1864
Hichler, George, Private	E	Gettysburg	July 3 1863
Hicks, George H., Corporal	I	Smithfield	August 29 1864
Higgins, Charles W., Private	D	Cedar Creek	October 19 1864
Hill, Philip H., Corporal	E	Gettysburg	July 3 1863
Hirner, Louis, Private	I	Yellow Tavern	May 11 1864
Hobbs, Levant, Sergeant	C	Unknown	Died June 6 1864
Hodge, Milton, Private	K	Brandy Station	October 11 1863
Huff, John A., Private	E	Cold Harbor	June 1 1864
Jackson, Andrew T., Private	A	Brandy Station	October 11 1863
James, Aaron B., Sergeant	H	Newtown	November 12 1864
Johnson, Julius C., Private	D	Newtown	November 12 1864
Kennedy, Philip, Private	H	Front Royal	August 16 1864

Kennicut, James C., Private	I	Berryville by Guerrillas	August 19 1864
Kent, Francis P., Private	G	Gettysburg	July 3 1863
Lewis, Eaton, Private	M	Berryville by Guerrillas	August 19 1864
Little, John M., Private	M	Brandy Station	October 11 1863
Lusk, John F., Sergeant	K	Winchester	October 19 1864
Lutz, John G., Private	C	Berryville by Guerrillas	August 19 1864
McChusen, J. B., Private	G	Buckland Mills	October 19 1863
McCormick, William J., Private	D	Dinwiddie Courthouse	April 1 1865
McCrary, Calvin, Private	M	Haw's Shop	May 28 1864
McGuire, John, Private	F	Appomattox	April 8 1865
McIntyre, John D., Private	C	Brandy Station	October 12 1863
Maguire, Christopher, Private	I	Yellow Tavern	May 12 1864
Mann, Harvey W., Corporal	I	Shepherdstown	August 25 1864
Marshall, Norton C., Sergeant	I	Hanovertown (?)	May 27 1864
Mather, Zelotes H., Sergeant	M	Boonesborough	July 8 1863
Meyer, George W., Private	M	Luray	September 24 1864
Miller, Daniel F., Sergeant	L	Unknown	Died June 14 1864
Mills, James F., Private	M	Richmond	March 1 1864
Morgan, Isaac C., Private	E	Newtown	November 12 1864
North, William O., Captain	F	Winchester	September 19 1864

Notting, John, Private	I	Gettysburg	July 3 1863
O'Brien, Anthony, Private	A	Yellow Tavern	May 11 1864
O'Brien, John, Private	A	By Guerrillas	December 2 1864
O'Brien, Matthew, Private	A	Loudon County	November 1864
Olaphant, David, Captain	B	Haw's Shop	May 28 1864
Osborn, Isaac C., Private	M	Berryville by Guerrillas	August 191864
Perkins, Isaac, Private	A	Smithfield	August 291864
Phelan, Thomas, Sergeant	L	Cedar Creek	October 19 1864
Phillips, Edward H., Sergeant	H	Trevilian Station	June 11 1864
Prouty, Wallace, Private	E	Newtown	November 23 1864
Purdy, Robert, Private	H	Trevilian Station	June 12 1864
Ragan, Alexander, Corporal	C	Appomattox	March 31 1865
Rathburn, Chauncey J., Private	D	Hanover	June 30 1863
Reed, Arthur, Private	H	Haw's Shop	May 28 1864
Roberts, Ephraim, Sergeant	E	Cedar Creek	October 19 1864
Rockwell, Floyd, Private	L	Ashby's Gap	July 21 1863
Rockwell, William H., Corporal	I	Brandy Station	October 11 1863
Roe, Alva, Private	B	Smithfield	August 291864
Russell, Major W., Private	M	Summit Point	September 5 1864
Ryan, Michael, Private	K	Gettysburg	July 3 1863

Ryder, Stephen, Sergeant	D	Haw's Shop	May 28 1864
Scates, Charles, Private	A	Trevilian Station	June 11 1864
Shafer, Absalom B., Private	C	Berryville by Guerrillas	August 191864
Shattuck, Oscar, Private	K	Boonesborough	July 8 1863
Shrontz, Mortimer J., Sergeant	M	Smithburg	July 5 1863
Sickman, Simon, Private	F	James City	November 11863
Skeels, Squire E., Sergeant	M	Berryville by Guerrillas	August 191864
Skinner, Irwin M., Private	G	Gettysburg	July 3 1863
Smith, Joseph W., Private	H	Trevilian Station	June 11 1864
Smith, Stephen, Private	B	Brandy Station	October 11 1863
Spencer, Lucien H., Private	A	Trevilian Station	June 11 1864
Sprague, Almerin, Private	H	Haw's Shop	May 28 1864
Stewart, Harrison C., Private	E	Trevilian Station	June 11 1864
Taggart, Robert G., Private	H	Winchester	September 19 1864
Taylor, David P., Corporal	I	By Accident	March 27 1863
Tenney, Wayland, Corporal	H	Winchester	September 19 1864
Todd, Andrew, Private	F	Newtown	November 12 1864
Tuller, Calvin, H., Private	H	Shepherdstown	August 251864
Van Bree, Garrett, Private	L	Trevilian Station	June 11 1864
Vicory, William L., Private	M	Smithfield	August 291864
Warner, Oliver M.,	C	Berryville by	August

Private		Guerrillas	191864
Warren, Milan S., Private	K	Newby's Crossroads	July 24 1863
Watkins, Jarius, Corporal	M	Wilderness	May 6 1864
Wire, George, Private	M	Berryville by Guerrillas	August 191864
Whirehead, Richard H., Sergeant	A	Haw's Shop	May 28 1864
Withington, Hiram A., Corporal	M	Berryville by Guerrillas	August 191864
Wixsom, George, Private	I	Trevilian Station	June 11 1864
Wood, Edwin W., Sergeant	A	Shepherdstown	August 251864
Wood, Fletcher, Sergeant	A	Cedar Creek	October 19 1864
Wright, Francis M., Corporal	M	Unknown	September 10 1864
Yoek, George, Private	E	Yellow Tavern	May 11 1864

SIXTH MICHIGAN CAVALRY

NAME AND POSITION	COMPANY	BATTLE	DATE OF DEATH
Andrus, James L., Sergeant	H	Smithfield	August 291864
Avery, Marvin E., Sergeant	E	Trevilian Station	June 12 1864
Bacon, Truman J., Private	F	Falling Waters	June 14 1863
Barber, George, Private	E	Trevilian Station	June 11 1864
Barnes, Augustus M., Private	D	On Sultana Explosion	April 26 1865
Barnum, Andrew, Private	A	Winchester	September 19 1864

Bass, Nathan B., Private	E	Woodstock	October 81864
Batson, Charles, Private	B	Falling Waters	July 14 1863
Beckwith, George, Private	C	Trevilian Station	June 12 1864
Bolza, Charles E., Lieutenant	B	Falling Waters	July 14 1863
Bowman, Lewis, Private	B	Battle Mountain	July 24 1863
Briggs, George, Private	F	Cedar Creek	October 19 1864
Brockway, William F., Corporal	H	Cedar Creek	October 19 1864
Brockway, William M., Private	H	Rapidan River	September 16 1863
Brown, George F., Private	I	Trevilian Station	June 11 1864
Brown, James W., Sergeant	E	Haw's Shop	May 28 1864
Brownell, George H., Private	K	Gettysburg	July 3 1863
Buck, Charles, H., Private	D	Haw's Shop	May 28 1864
Burden, John, Private	B	Cedar Creek	October 19 1864
Burns, James, Private	D	Trevilian Station	June 12 1864
Butler, Edward, Private	M	Trevilian Station	June 11 1864
Campbell, Duncan, Private	M	Haw's Shop	May 28 1864
Campbell, William P., Private	C	Winchester	September 19 1864
Carey, Seth, Corporal	E	Haw's Shop	May 28 1864
Curliss, William, Private	G	Falling Waters	July 14 1863
Chandler, George B., Private	D	Somerville Ford	September 16 1863
Chase, Albert,	L	Somerville Ford	September 16

Teamster			1863
Clark, George, Corporal	G	Boonesborough	July 11 1863
Clark, Joshua P., Private	F	Muddy Branch	July 21 1864
Cole, Osmer F., Captain	G	By Indians	August 311865
Coon, Alexander H., Private	A	Trevilian Station	June 11 1864
Cox, Charles W., Private	C	Hunterstown	July 2 1863
Cranston, Thomas C., Private	C	Haw's Shop	May 28 1864
Cryderman, John, Private	E	Yellow Tavern	May 11 1864
Daily, William H., Sergeant	D	Hanovertown	May 27 1864
Day, John, Private	F	Washington	April 27 1865
Decker, Almeron, Private	E	Yellow Tavern	May 11 1864
Dexter, Dallas, Private	M	Waterford	August 8 1863
Dixon, William G., Private	C	Buckland Mills	October 19 1863
Dudley, Jerry, Private,	I	Cedar Creek	October 19 1864
Earl, Benjamin F., Private	I	Somerville Ford	September 16 1863
Edie, Thomas A., Lieutenant	A	Meadow Bridge	May 12 1864
Edwards, William H., Private	E	Yellow Tavern	May 11 1864
Eldridge, Marvin J., Private	C	Haw's Shop	May 28 1864
Elmore, Byron A., Private	B	Haw's Shop	May 28 1864
Fairbanks, Forrest, Private	D	Trevilian Station	June 12 1864
Farrell, Thomas, Private	M	Harper's Ferry	August 171863

Fay, George W., Private	H	Trevilian Station	June 12 1864
Finney, Solon H., Lieutenant	E	Beaver Mills	April 4 1865
Foe, James, Private	C	Hatcher's Mills	April 4 1865
Foote, Martin W., Private	C	Trevilian Station	June 11 1864
Francisco, James K., Sergeant	K	Winchester	September 19 1864
Galusha, Sears E., Corporal	G	Falling Waters	July 14 1863
Gooch, Horace N., Private	B	Falling Waters	July 14 1863
Griffith, Gilbert D., Private	B	Falling Waters	July 14 1863
Gross, Charles H., Private	M	Haw's Shop	May 28 1864
Hanna, John, Private	A	Haw's Shop	May 28 1864
Harding, Ira C., Private	D	Somerville Ford	September 16 1863
Harrison, Henry M., Private	E	High Bridge	April 6 1865
Hart, Horace, Corporal	D	Hanover	June 30 1863
Hawkins, Oscar J., Private	K	Trevilian Station	June 11 1864
Hayes, William O., Private	C	By Indians	September 13 1865
House, Martin, Private	M	Haw's Shop	May 28 1864
Hughson, Franklin, Private	K	Trevilian Station	June 11 1864
Hulet, James H., Private	K	On Sultana Explosion	April 26 1865
Hutchinson, Miles E., Private	E	Haw's Shop	May 28 1864
Ingersoll, George B. W., Sergeant	G	Shepherdstown	August 281864
Inman, Elisha,	K	By Guerrillas	December 4

Private			1864
Jewell, Leander, Sergeant	A	Hanovertown	May 27 1864
Jewett, Aaron C., Acting Adjutant		Williamsport	July 6 1863
Johnson, Warren E., Private	I	Seneca	June 11 1863
Johnson, William W., Private	M	Unknown	October 11 1864
Jolly, Toussaint, Private	I	Haw's Shop	May 28 1864
Jones, Levi F., Private	D	Haw's Shop	May 28 1864
Kelsey, Ira, Private	K	Newby's Crossroads	July 24 1863
Kilbourn, Joseph, Private	C	Winchester	September 19 1864
Kirkby, Henry, Private	I	Haw's Shop	May 28 1864
Krauss, Charles C., Private	A	Hunterstown	July 2 1863
Larime, Joseph, Private	C	Wilderness	May 6 1864
Livingston, Albert, Private	C	Thornton Gap	July 24 1863
Livingston, Monroe, Private	F	Falling Waters	July 14 1863
Lorsey, Charles, Private	B	Falling Waters	July 14 1863
Lyons, James, Private	F	Unknown	July 30 1864
McClure, Alexander, Private	C	Haw's Shop	May 28 1864
McDonald, Jeremiah, Private	F	Falling Waters	July 14 1863
McLean, Peter, Corporal	G	Haw's Shop	May 28 1864
Martin, Alonzo R., Private	B	Falling Waters	July 14 1863
Mathers, James, Captain	L	Winchester	August 11 1864

Mayfield, Oakland W., Private	B	Falling Waters	July 14 1863
Morre, Ezra P., Private	A	Trevilian Station	June 12 1864
Mosher, Merritt, Corporal	A	Wilderness	May 6 1864
Morrison, Edwin M., Private	K	Wilderness	May 6 1864
Moulthrop, Albert, Private	I	Tom's Brook	October 91864
Neal, Flavius J., Private	B	Falling Waters	July 14 1863
Nellins, John, Private	H	Winchester	November 18 1864
Onweller, William, Private	B	Trevilian Station	June 11 1864
Otis, Albert, Private	D	Falling Waters	July 14 1863
Patten, George T., Sergeant	B	Falling Waters	July 14 1863
Pelton, Francis, Private	B	Falling Waters	July 14 1863
Perkins, William, Private	E	Beaver Pond Mills	April 4 1865
Pixley, Austin, Private	A	Drowned	June 15 1864
Potter, Harvey B., Sergeant	B	Falling Waters	July 14 1863
Powers, Wesley, Private	I	Seneca	June 11 1863
Pray, Stephen, Private	C	Trevilian Station	June 12 1864
Provin, James J., Private	M	Smithfield	February 5 1864
Rappelye, Mortimer, Sergeant	C	Hanovertown	May 27 1864
Rodder, John, Private	I	Trevilian Station	June 11 1864
Richardson, Francis D., Private	F	Falling Waters	July 14 1863
Rider, Carlos,	D	Trevilian Station	June 11

Corporal			1864
Rogers, Frederick V., Private	G	Haw's Shop	May 28 1864
Rogers, Remus, Private	B	Falling Waters	July 14 1863
Roney, Charles E., Private	C	Dinwiddie Courthouse	April 9 1865
Rossell, Abram, Private	D	Falling Waters	July 14 1863
Royce, David G., Captain	D	Falling Waters	July 14 1863
Ruckel, George, Private	I	Trevilian Station	June 11 1864
Sliter, Josiah T., Private	B	Falling Waters	July 14 1863
Smith, Jonathan W., Private	K	Newby's Crossroads	July 24 1863
Soule, John W., Corporal	D	Boonesborough	July 8 1863
Stafford, Ananias, Private	D	Haw's Shop	May 28 1864
Stanton, Andrew, Private	K	Winchester	September 19 1864
Stowe, Stephen L., Sergeant	B	Trevilian Station	June 11 1864
Streeter, Seth, Private	H	Unknown	Died August 2 1863
Sweet, Lorenzo D., Private	I	Falling Waters	July 14 1863
Telling, George, Corporal	D	Boonesborough	July 8 1863
Trager, George, Private	F	Falling Waters	July 14 1863
Tucker, Ephraim, Private	D	Cedar Creek	October 19 1864
Tucker, Harvey, Sergeant	C	Wilderness	May 6 1864
Tuttle, Milo, Private	M	Waterford	August 8 1863
Von Helmerich, Frederick, Private	I	Seneca	June 11 1863

NAME AND POSITION	COMPANY	BATTLE	DATE OF DEATH
Wadeweitz, Frederick, Private	I	Meadow Bridge	May 12 1864
Ward, Erastus E., Private	F	Five Forks	April 1 1865
Weber, Peter A., Captain	B	Falling Waters	July 14 1863
Whalen, David, Private	I	Seneca	June 11 1863
Wheaton, Henry F., Private	H	Unknown	February 2 1865
White, William C., Private	D	Trevilian Station	June 11 1864
Wightman, George H., Sergeant	L	Unknown	September 1 1864
Williams, Edward L., Sergeant	I	Cedar Creek	October 19 1864
Williams, John D., Corporal	I	Trevilian Station	June 11 1864
Winters, John, Private	H	Accident	July 28 1864
Yax, John, Corporal	C	Cold Harbor	June 1 1864
Yeoman, Lewis H., Private	E	Brandy Station	October 11 1863

SEVENTH MICHIGAN CAVALRY

NAME AND POSITION	COMPANY	BATTLE	DATE OF DEATH
Adams, Oscar H., Corporal	A	Trevilian Station	June 12 1864
Adams, William H., Private	D	Gettysburg	July 3 1863
Armstrong, Harrison, Private	F	Yellow Tavern	May 11 1864
Baker, George, Private	B	Killed by Indians	August 5 1865
Bedel, Harlin, Corporal	F	Five Forks	April 1 1865
Bedel, James T.,	F	Gettysburg	July 3

Private			1863
Bouchard (Bershall), Eli, Private	K	Front Royal	August 161864
Brewer, Melvin, Lieutenant Colonel		Winchester	September 19 1864
Brickwell, Edward J., Private	A	Gettysburg	July 3 1863
Brownell, Horace R., Private	A	Gettysburg	July 3 1863
Bush, Christian, Corporal	D	Winchester	September 19 1864
Bush, Frederick, Corporal	D	Haw's Shop	May 28 1864
Carver, Lucius, Lieutenant	M	Front Royal	August 161864
Chapman, Frank, Private	A	Richmond	March 1 1864
Cheesman, Jeremiah, Private	F	Trevilian Station	June 11 1864
Church, Benjamin, Sergeant	C	Gettysburg	July 3 1863
Churchill, Alfred W., Corporal	G	Cedar Creek	October 19 1864
Clark, Edgar A., Private	A	By Accident	July 5 1865
Clark, Jonas, Private	K	Richmond	March 1 1864
Cochran, Harlan B., Sergeant	F	Falling Waters	July 14 1863
Cochran, William J., Corporal	I	Front Royal	August 161864
Campau, Peter, Private	D	Boonesborough	July1863
Cook, Elliott A., Sergeant	C	Robinson River	October 81863
Cooper, Eugene, Private	F	Trevilian Station	June 11 1864
Cornell, Llewellyn C., Private	B	Gettysburg	July 3 1863
Crampton, P. H.,	G	Hagarstown	July 6

Name	Company	Battle	Date
Private			1863
Croman, William, Private	E	Brandy Station	October 11 1863
Dann, Daniel, Private	K	Yellow Tavern	May 11 1864
Diehl, Henry, Private	C	Salem Church	June 2 1864
Delamater, Martin R., Corporal	G	Cold Harbor	June 1 1864
Dumphrey, Edwin, Sergeant	A	Winchester	September 19 1864
Edwards, William, Private	C	Trevilian Station	June 11 1864
Filbern, Owen, Private	I	Buckland Mills	October 19 1863
Finch, Robert, Private	E	Gettysburg	July 3 1863
Firman, Josiah B., Corporal	H	Yellow Tavern	May 11 1864
Fisher, Mathias, Bugler	B	Berryville	September 4 1864
Fordham, Albert, Corporal	D	Gettysburg	July 3 1863
Fox, William H., Corporal	M	Winchester	September 19 1864
Granger, Henry W., Major		Yellow Tavern	May 11 1864
Guio, Henry, Corporal	F	Falling Waters	July 14 1863
Haines, Henry, Private	D	Gettysburg	July 3 1863
Hall (Hull), William, Private	M	Buckland Mills	October 19 1863
Hamel, Harrison, Private	K	Winchester	September 19 1864
Haskins, James, Sergeant	B	Duck Pond Mills	April 3 1865
Hassart, Andrew, Private	B	Winchester	September 19 1864
Hasty, Robert, Private	I	Gettysburg	July 3 1863

Heinck, John, Saddler	A	Gettysburg	July 3 1863
Hoag, Robert, Private	F	Gettysburg	July 3 1863
Hopkins, Horace, Private	E	Gettysburg	July 3 1863
House, Barnum B., Private	E	Old Church	May 23 1864
Jackson, Orlando D., Private	D	Gettysburg	July 3 1863
Jessup, Charles H., Private	F	Trevilian Station	June 11 1864
Karcher, Jehial, Private	D	Gettysburg	July 3 1863
Keller, Henry H., Private	B	Todd's Tavern	May 7 1864
Kisner, Samuel, Private	C	By Accident	July 18 1863
Koster, Frederick, Private	H	Cedar Creek	October 19 1864
Laird, William J., Sergeant	B	Cedar Creek	October 19 1864
Lake, John W., Private	A	Haw's Shop	May 28 1864
Larrue, Hiram J., Private	B	By Guerrillas	March 28 1864
Long, Edward, Private	B	Winchester	September 19 1864
Lundy, George W., Private	H	Gettysburg	July 3 1863
McClure, Ralph, Private	H	Yellow Tavern	May 11 1864
McComber, William, Private	C	Cold Harbor	June 4 1864
McDonald, John J., Sergeant	C	Gettysburg	July 3 1863
McLaine, Alexander, Private	E	Marselas	May 22 1863
Martin, Francis D., Private	H	Gettysburg	July 3 1863
Matchett, Noel,	A	Trevilian Station	June 11

			1864
Private			1864
Mead, Joseph L., Lieutenant	L	Smithfield	August 291864
Mercer, Thomas, Private	F	Smithfield	August 291864
Milbourn, John L., Corporal	D	Gettysburg	July 3 1863
Miller, Jacob L., Private	C	Unknown	Died June 21 1864
Mills, Harry, Private	H	Yellow Tavern	May 11 1864
Miner, Charles E., Sergeant	F	Gettysburg	July 3 1863
Moll, Cornelius, Private	F	White Ford	September 22 1863
Motley, Thomas, Private	G	Gettysburg	July 3 1863
Nay, Harmon, Private	E	Hagarstown	July 6 1863
Nichols, William H., Private	H	Trevilian Station	June 11 1864
Nolan, Arthur D., Sergeant	I	Haw's Shop	May 28 1864
O'Brien, William H., Sergeant	A	Yellow Tavern	May 11 1864
Olin, Oscar O., Private	M	Cedar Creek	October 19 1864
Page, Truman, Bugler	F	Yellow Tavern	May 11 1864
Parks, Allen C., Private	A	Cedar Creek	October 19 1864
Paule, Jacob, Sergeant	F	Yellow Tavern	May 11 1864
Perkins, Myron H., Sergeant	B	Haw's Shop	May 28 1864
Ploof, Dewitt C., Private	K	Trevilian Station	June 11 1864
Pomeroy, David H., Private	L	Trumble Run	June 9 1864
Ralph, Oscar S., Corporal	F	Falling Waters	July 14 1863

Richards, William H., Private	H	Emmittsburg	July 4 1863
Robinson, James B., Teamster	E	Tom's Brook	October 91864
Shafer, Charles F., Private	A	Winchester	September 19 1864
Smith, Alonzo, Private	C	Gettysburg	July 3 1863
Smith, Eli, Private	K	Gettysburg	July 3 1863
Smith, Perry W., Private	H	Hagarstown	July 13 1863
Spear, Truman, Private	G	Gettysburg	July 3 1863
Stearns, William A., Sergeant	B	Cold Harbor	June 1 1864
Stephens, Charles, Private	K	Front Royal	August 161864
Taber, Winfield S., Sergeant	M	Culpeper	September 13 1863
Thompson, Henry, Private	D	Smithfield	August 291864
Treat, Gordon, Private	K	Front Royal	August 161864
Vancourse, Henry, Private	K	Front Royal	August 161864
Van Duzer, Charles E., Private	M	Unknown	September 1864
Van Ness, George E., Corporal	M	Gettysburg	July 3 1863
Walters, Nelson, Private	A	Gettysburg	July 3 1863
Whittaker, William S., Private	B	Trevilian Station	June 11 1864
Wilcox, Charles, Corporal	A	Gettysburg	July 3 1863
Williams, Squire, Corporal	I	Haw's Shop	May 28 1864

TOTAL NUMBER KILLED IN ACTION

First Michigan Cavalry	157
Fifth Michigan Cavalry	144
Sixth Michigan Cavalry	141
Seventh Michigan Cavalry	106
Total	**548**

FOOTNOTES

[Footnote 1: Quoted from "Michigan in the War."]

[Footnote 2: The original roster of the regiment may be found in appendix "A" to this volume.]

[Footnote 3: Grand Rapids, Michigan, so named on account of its location in the heart of the valley of Grand River. Also known as the "Furniture City," referring to its chief industry.]

[Footnote 4: Robert Williams, a Virginian, grandson of James Williams of the Virginia line in the Revolution. He married the widow of Stephen A. Douglas.]

[Footnote 5: Third Michigan Infantry. It served three years, and was then reorganized as the "New Third."]

[Footnote 6: Since the above was written, I have become satisfied that this man was really taken prisoner and that he died as such in the Confederate prison at Andersonville. His name appears on one of the markers in the national cemetery there.]

[Footnote 7: September, 1907.]

[Footnote 8: Official Records, Series 1, Vol. XXVII, Part III, page 276.]

[Footnote 9: The Michigan Cavalry Brigade was the outgrowth of the reorganization of the Federal cavalry that followed Lee's invasion of the North and Hooker's consequent movement into Maryland. It consisted originally, as has been shown, of three regiments - the Fifth, Sixth and Seventh. They were all organized in 1862, spent the winter of 1862-63 in camp on Meridian and

Capitol Hills, Washington, D. C., and during the spring months of the latter year, were engaged in doing outpost duty in Fairfax County, Va., within the defenses of Washington. They were, therefore, in the language of another, "fresh from pastures green" when General Hooker, en route to Maryland in June, 1863, picked them up in passing and made them a part of that grand Army of the Potomac which, on the battlefield of Gettysburg, won a renown as lasting as history itself.

The commanding officer was Brigadier General J.T. Copeland, a Michigan man, promoted from the colonelcy of the Fifth. The battalion commanders were, respectively, Colonels Russell A. Alger, George Gray and William D. Mann. The first had seen service in the Second Michigan as captain and major, under Colonels Gordon Granger and P.H. Sheridan; the last in the First Michigan, under Brodhead and Town. Colonel Gray was appointed from civil life, and was having his first experience of "war's rude alarums."]

[Footnote 10: Custer in his report mistook the York for the Hanover road.]

[Footnote 11: General Custer mistook the Low Dutch for the Oxford road.]

[Footnote 12: A letter from General Gregg to the writer says: "There is no conflict between your recollection and mine as to the events of that day."-- J.H.K.]

[Footnote 13: A possible solution of this difficulty has come to my mind. It is this. That Custer originally wrote "1 o'clock" and that in copying the "1" and the "o" were mistaken for "10," and o'clock added--J.H.K.]

[Footnote 14: In this connection it may be stated that Colonel Fox's history of the casualties in the war shows that there were 260 cavalry regiments in the service of the Union. Of these, the First Michigan lost the largest number of men killed in action of all save one - the First Maine. In percentage of killed, in proportion to numbers the Fifth and Sixth Michigan rank all the rest, not excepting the two first named, and it must be remembered that the Fifth and Sixth went out in 1862 and did their first fighting in the Gettysburg campaign. They stand third and fourth in the number killed, being ranked in that respect by the First Maine and First Michigan alone. The four regiments in the Michigan Brigade during their terms of service lost twenty-three officers and 328 men killed; eight officers and 111 men died of wounds; nine officers and 991 men died of disease - a grand total of 1470 officers and men who gave up their lives during those four years of war. - J.H.K.]

[Footnote 15: It may be proper to state that during the Gettysburg campaign the Michigan Brigade lost thirty officers killed and wounded, whose names are here given.

KILLED
First Michigan - Capt. W.R. Elliott, Capt. C.J. Snyder, Lieut. J.S. McElhenny - 3.
Fifth Michigan - Major N.H. Ferry - 1.
Sixth Michigan - Major P.A. Weber, Capt. D.G. Royce, Lieut. C.E. Bolza, Acting Adjutant A.C. Jewett - 4.

WOUNDED
First Michigan - Capt. D.W. Clemmer, Lieut. E.F. Baker, Capt. A.W. Duggan, Capt. George W. Alexander, Capt. H.E. Hascall, Capt. W.M. Heazlett, Capt. G.R. Maxwell, Lieut. R.N. Van Atter - 8.
Fifth Michigan - Col. R.A. Alger, Lieut. Col. E. Gould, Lieut. T. Dean, Lieut. G.N. Dutcher - 4.
Sixth Michigan - Lieut. George W. Crawford; Capt. H.E. Thompson, Capt. J. H. Kidd, Lieut. E. Potter, Lieut. S. Shipman - 5.
Seventh Michigan - Lieut. J.G. Birney, Lieut. J.L. Carpenter, Lieut. E. Gray, Lieut. C. Griffith, Capt. Alex. Walker - 5.]

[Footnote 16: Brigadier General Henry E. Davies, formerly colonel Second New York Cavalry, assigned as permanent successor of Farnsworth, killed at Gettysburg.]

[Footnote 17: Attached to the Michigan Brigade.]

[Footnote 18: Rosser, Young and Gordon.]

[Footnote 19: Since reporting for duty, October 12, I had been in command of the regiment.]

[Footnote 20: Campaigns of Stuart's Cavalry.]

[Footnote 21: Fitzhugh Lee was Custer's instructor in West Point before the war broke out.]

[Footnote 22: Kilpatrick's Report, Official Records, series I. vol. XXXIII. p. 133.]

[Footnote 23: "Unless the separate commands in an expedition of this nature are very prompt in movement, and each equal to overcoming at once any obstacle it may meet combinations rarely work out as expected."--Personal Memoirs of P.H. Sheridan, vol. I, p. 373.]

[Footnote 24: A small stream crossing the turnpike and after which the historical pike was named.]

[Footnote 25: On page 813, Vol. XXXVI, Series I, Part 1, of the War Records, in the report of General Merritt appears the following: "A charge made, mounted, by one regiment of the First Brigade, (the Fifth Michigan)." The words in parenthesis should be the First Michigan. It is a pity that the official records should thus falsify history.]

[Footnote 26: I am not positive that these were the particular tunes the bands played.]

[Footnote 27: Personal Memoirs of P.H. Sheridan, Vol. I: page 417. Also Records, Series I, Vol. XXXVI, part 1.]

[Footnote 28: Taken from the official records in the office of the adjutant general of Michigan, Lansing, Michigan.]

[Footnote 29: Taken from the official records in the office of the adjutant general of Michigan, Lansing, Michigan.]

[Footnote 30: Official Records, Series I, Vol. XXXVI, part I, page 851.]

[Footnote 31: Official Records, Series I, Vol. XXXVI, part I, page 810.]

[Footnote 32:
Headquarters 1st Brig. 1st Div. Cavalry Corps, June 3, 1864
To His Excellency
Governor Blair,
I most cheerfully and earnestly recommend that the foregoing petition may be granted. Major Kidd has commanded his regiment for several months. He has distinguished himself in nearly all of the late severe engagements of the corps. Michigan cannot boast of a more gallant or efficient officer than Major Kidd, and I am confident that his appointment as colonel of the 6th would not only produce entire satisfaction in his regiment, but would serve to increase the already high

but well earned fame of the Michigan Cavalry Brigade.

Very respectfully, etc.,

G.A. CUSTER,

Brig. Gen'l Comdg.]

[Footnote 33: The title given to the department over which Sheridan was to have supreme command, and which included West Virginia.]

[Footnote 34: Torbert had been created "chief of cavalry," and Merritt assigned to command of the First Division. Colonel Charles R. Lowell, Second Massachusetts cavalry succeeded Merritt in command of the Reserve Brigade.]

[Footnote 35: Attached temporarily to the Michigan Brigade.]

[Footnote 36: Second division of cavalry from West Virginia, General W.W. Averell.]

[Footnote 37: Written in 1886.]

[Footnote 38: The only order I had received at the time was to support the picket line with the entire brigade. See General Merritt's report, Official records, Vol. XLIII, series I, part I, page 449.]

[Footnote 39: The Fifth United States cavalry, General Merritt's escort. General Merritt's report.]

[Footnote 40: General Sheridan's report states that it was Getty's division of the Sixth corps only that was in this position when he came up - that the other divisions were farther to the rear but were brought up to the alignment.

"On arriving at the front, I found Merritt's and Custer's divisions of cavalry, * * * and Getty's division of the Sixth Corps opposing the enemy. I suggested to General Wright that we would fight on Getty's line, and that the remaining two divisions of the Sixth Corps, which were to the right and rear about two miles, should be ordered up, * * before the enemy attacked Getty." - Sheridan's report, Records, Vol. XLIII, part I, page 53.]

[Footnote 41: "The First brigade, in column of Regiments in line, moved forward like an immense wave, slowly at first, but gaining strength and speed as it progressed, overwhelmed a battery and its supports amidst a devastating shower of canister and a deadly fire of musketry from part of Kershaw's division, at short range from a heavy wood to our left. Never has the mettle of

445

the division been put to a severer test than at this time, and never did it stand the test better. The charge was made on an enemy well formed and prepared to receive it with guns double-shotted with canister." - General Merritt's official report, Records, Vol. XLIII, Part I, page 450.]

[Footnote 42: Records, Series I, Vol. XLIII, part I, page 136.]

[Footnote 43: Records. Series I, Vol. XLIII, part I. pages 136-37.]

[Footnote 44: Records. Series I, Vol. XLIII, part I. page 453.]

ILLUSTRATIONS CREDITS

COVER: Gettysburg dead, New York Historical Society; Sunrise cannon, C.Kurt Holter/Shutterstock.com. Art and graphics by Nikki D. Cirignani

A NATIONAL AWAKENING: University of Michigan, University of Michigan Photographs Vertical File; Detroit, University of Michigan Photographs Vertical File; Stephen Douglas, Library of Congress; William Lloyd Garrison, National Archives; Tappan speech, University of Michigan Photographs Vertical File

AN EVENTFUL WINTER: Andrew White, University of Michigan; Tom Weir, Library of Congress

RECRUITING IN MICHIGAN: Detroit, Detroit Public Library; Lewis Cass home, Detroit Public Library; Lewis Cass, Library of Congress

THE SUMMER OF 1862: Richmond, Library of Congress; McClellan and staff, Library of Congress

JOINING THE CAVALRY: Michigan logging, Burton Historical Collection of the Detroit Public Library; Wilder Foster, National Archives

IN THE REGIMENTAL RENDEZVOUS: Winter camp, Library of Congress; Custer, from original publication of subject book; Russell Alger, original publication of subject book; Champlain Greeley, Courtesy John Dickey; Commissary tent, Library of Congress; Williamsport, Library of Congress; Wagon train, Library of Congress

THE DEPARTURE FOR WASHINGTON: Michigan Central Railroad Station, Detroit Public Library

THE ARRIVAL IN WASHINGTON: Mount Pleasant (Harewood) Hospital, Library of Congress; Capital dome, Library of Congress; Troops in drill, Library of Congress

THE STAY IN WASHINGTON: General Heintzelman, National Archives; Armory Square Hospital, Library of Congress

FIELD SERVICE IN VIRGINIA: Washington DC, Library of Congress; Alexandria, Scene of Ellsworth killing, Library of Congress; Troops in camp, Library of Congress; Fairfax Court House, Library of Congress; Vienna, cavalry camp, Library of Congress, National Hotel, Gettysburg Daily; Chain Bridge, Library of Congress

IN THE GETTYSBURG CAMPAIGN: Troops, Library of Congress; Mount St. Mary's, Library of Congress, Gettysburg Cemetery Gate, Library of Congress; Pennington and battery, Library of Congress; Hanover Junction Railway Station, Library of Congress; Custer, original Kidd publication; Hunterstown, The Story of the Battle of Gettysburg and the Field, 1913; General Gregg and staff, National Archives; Major Trowbridge, Library of Congress; Gettysburg, Library of Congress

FROM GETTYSBURG TO FALLING WATERS: John Elon Farnsworth, Library of Congress; General Kilpatrick, Library of Congress; South Mountain, Library of Congress; Confederate dead by a road near Hagerstown, Library of Congress; Boonsborough, Library of Congress; Peter Weber, original Kidd publication

FROM FALLING WATERS TO BUCKLAND MILLS: Field hospital, Library of Congress; Ambulance train, Library of Congress, Washington House, Vendome Hotel Postcard pre-1923; Senator Henry Wilson, Library of Congress; Brandy Station, Library of Congress; Brandy Station RR station, Library of Congress; Libby Prison, Library of Congress, Custer and officers, National Park Service

THE BATTLE OF BUCKLAND MILLS: Warrenton Pike, Library of Congress; Fitzhugh Lee, Library of Congress

WINTER QUARTERS IN STEVENSBURG: Winter camp, National Archives; Jacob Howard, National Archives; Thorton Stringfellow, Kansas Historical Society; Custer Wedding, Library of Congress; Culpepper, Library of Congress; Wesley Merritt, Library of Congress; Richmond, Library of Congress; Jamestown Exposition, Virginia Historical Society

THE WILDERNESS CAMPAIGN: Sheridan and officers, Library of Congress; Chancellorsville, Library of Congress; Todd's Tavern, Library of Congress

THE YELLOW TAVERN CAMPAIGN: Spottsylvania Court House, Library of Congress; Grant and staff, Library of Congress; Telegraph Road, Library of Congress; Angelo Tower, original Kidd publication; Stuart's grave, New York Historical Society

YELLOW TAVERN TO CHESTERFIELD STATION: Haxall's Landing, Library of Congress; Damaged bridge over Chickahominy, Mathew Brady Collection; Gaines Mill, Library of Congress; Destroyed train near Richmond, Library of Congress

HANOVERTOWN AND HAW'S SHOP: Wade Hampton, Library of Congress; Cold Harbor Battlefield, Library of Congress

THE TREVILIAN RAID: Manning Birge, original Kidd publication; King and Queen County Court House, photo by Louis Shepherd

IN THE SHENANDOAH VALLEY: Harper's Ferry, DeGolyer Library, Southern Methodist University; Confederate prisoners taken at Fisher's Hill, Library of Congress; Front Royal, National Archives; Mosby and battalion, Library of Congress; Shepherdstown, ruins of bridge, Library of Congress; Opequon Creek, Library of Congress; Mill on Santa Anna, National Archives

THE BATTLE OF CEDAR CREEK: Cedar Creek, Library of Congress; Cavalry Headquarters, Charles Lowell, Library of Congress; Cedar Creek Battlefield, Library of Congress; Charles Lowell, original Kidd publication

A MYSTERIOUS WITNESS: Grand Review, National Archives; John Wilkes Booth, Library of Congress
A MEETING WITH MOSBY: Mosby, Library of Congress, Spotswood Hotel, Library of Congress

A preview into *History in Words and Pictures Series,* Book Six

A COMPLETE LIFE OF GENERAL GEORGE A. CUSTER

MAJOR-GENERAL OF VOLUNTEERS; BREVET MAJOR-GENERAL, U.S. ARMY; AND LIEUTENANT-COLONEL, SEVENTH U.S. CAVALRY

By
Frederick Whittaker

CHAPTER 1 - EARLY LIFE

GEORGE ARMSTRONG CUSTER was born in New Rumley, Ohio, December 5, 1839. New Rumley is a group of houses, an old established settlement in Harrison County on the border of Pennsylvania, and peopled from thence early in the last century. It is a small place, not set down on any but very large scale maps, and most of the population of the township is scattered in farm houses about the country. The family history, gleaned from the family bible, is plain and simple. It is that of an honest group of hard workers, not ashamed of work, and it shows that the stock of which the future general came was good, such as made frontiersmen and pioneers in the last century.

Emmanuel H. Custer, father of the general, was born in Cryssoptown, Alleghany County, Maryland, December 10th, 1806. Today, a hale hearty old man of seventy, somewhat bowed but well as ever to all seeming, he stands a living instance of the strong physique and keen wits of the determined men who made the wild forests of Ohio to bloom like the rose. He was brought up as a smith and worked at his trade for many years, until he had saved enough money to buy a farm, when he became a cultivator. All he knows, he taught himself, but he gave his children the best education that could be obtained in those early days in Ohio.

When quite a young man, he left Maryland and settled in New Rumley, being the only smith for many miles. He prospered so well that he was able to get married when twenty-two years of age. He married Matilda Viers, August 7th, 1828, and their marriage lasted six years, during which time three children were born of whom only one, Brice W. Custer, of Columbus, Ohio, is now living. He is bridge inspector on one of the railroads leading from that place. The first Mrs. Custer died July 18th, 1834.

The maiden name of the second Mrs. Custer, mother of the general, was Maria Ward. She was born in Burgettstown, Pennsylvania, May 31st, 1807, and was first married when only a girl of sixteen to Mr. Israel R. Kirkpatrick. Her husband died in 1835, a year after the death of the first Mrs. Custer. The widow Kirkpatrick had then three children, whereof two are now alive. David Kirkpatrick lives in Wood County, Ohio, some forty miles south from Toledo. Lydia A. Kirkpatrick married Mr. David Reed of Monroe, Michigan, and in after she became more than a sister - a second mother, to the subject of our biography.

After two years widowhood, Mrs. Kirkpatrick married Emmanuel Custer, April 14th, 1837, and became the mother of the general two years later, as the second Mrs. Custer. She is still, at the present date of writing, living, but in very feeble health. The children of this second marriage were born as follows:

1. George Armstrong Custer, December 5, 1839
2. Nevin J. Custer, July 29, 1842
3. Thomas W. Custer, March 15, 1845
4. Boston Custer, October 31, 1848
5. Margaret Emma Custer, January 5, 1852

All were born in Harrison County, in or near New Rumley. Nevin and Margaret alone now survive, the latter the widow of Lieutenant Calhoun, who was killed on the field of battle with his three brothers-in-law, June 25th, 1876. Nevin Custer now lives on a farm near Monroe, Michigan. During the late war, he enlisted as a private soldier, but was thrown out for physical disability in spite of his anxiety to serve his country. He had all the spirit of the Custers, but lacked the good physique of the other members of the family.

I have been thus particular in giving the family record, because little is known to the world on that subject. It is the record of a plain yeoman family, such as constitutes the bone and sinew of the country. The name of Custer was originally Kuster, and the grandfather of Emmanuel Custer came from Germany; but Emmanuel's father was born in America.

The grandfather was one of those same Hessian officers over whom the colonists wasted so many curses in the Revolutionary war, and who were yet so innocent of harm and such patient, faithful soldiers. After Burgoyne's surrender in 1778, many of the paroled Hessians seized the opportunity to settle in the country they came to conquer; and amongst these the grandfather of Emmanuel Custer, captivated by the bright eyes of a frontier damsel, captivated her in turn with his flaxen hair and sturdy Saxon figure, and settled down in Pennsylvania; afterwards moving to Maryland.

It is something romantic and pleasing after all, that stubborn

George Guelph, in striving to conquer the colonies, should have given them the ancestor of George Custer, who was to become one of their greatest glories.

Of this family, the boy George Armstrong was born, and grew up a sturdy, flaxen-headed youngster, full of life and frolic, always in mischief and yet strange to say, of the gentlest and most lovable disposition. The closest inquiry fails to reveal a single instance of ill temper during Custer's boyhood. All his playmates speak of him as the most mischievous and frolicsome of boys, but never as quarrelsome. There is actually not a single record of a fight in all his school life, though the practical jokes are without number.

Continued in *History in Words and Pictures Series,*
Book Six:

A COMPLETE LIFE OF GENERAL GEORGE A. CUSTER
MAJOR-GENERAL OF VOLUNTEERS; BREVET MAJOR-GENERAL, U.S. ARMY; AND LIEUTENANT-COLONEL, SEVENTH U.S. CAVALRY
By Frederick Whittaker

www.ingramcontent.com/pod-product-compliance
Lightning Source LLC
Chambersburg PA
CBHW062355090426
42740CB00010B/1288